Mapping the Heavens

Mapping
the Heavens

Peter Whitfield

New Edition

First published in 1995
This edition published 2018 by
The British Library
96 Euston Road
London NW1 2DB

British Library Cataloguing in Publication Data
A catalogue record for this book is available from the British Library

ISBN 978 0 7123 5265 9
Picture Research by Sally Nicholls
Designed by Chris Benfield
Printed and bound in Malta by Gutenberg Press

Frontispiece:
Mattheus Seutter: The Comet of
1742 (see pages 136–37).

Contents

Preface

'The strongest affection and utmost zeal should, I think, promote the studies
concerned with the most beautiful objects. This is the discipline which deals
with the universe's divine revolutions, the stars' motions, sizes, distances, risings
and settings ... for what is more beautiful than heaven? By virtue of heaven's
transcendent perfection, most philosophers have called it a visible god... .
However this divine rather than human science is not free from perplexities ...'
    Copernicus: *On the Revolutions of the Heavenly Spheres*, 1543

This book traces the history of astronomy through its images, in particular through the development of celestial maps, where science and art came together in the attempt to shape a rational image of the heavens. To watch and interpret the skies has always been one of humanity's fundamental instincts. In all pre-industrial societies, real darkness filled a great part of people's lives, and ancient civilisations built up a knowledge of the skies that was in some ways more precise than their knowledge of the world in which they lived. To impose order on the expanse of star-filled sky, star groups in the form of animals, gods and heroes were created as zones or landmarks. With the heavens mapped in this way, the patterns of the night sky could be used as fundamental gauges of direction and, more importantly, of time, for no civilisation could progress without a calendar.

But alongside this precise, observed regularity there was always the mystery of the skies: not content merely to observe, astronomers sought to explain the causes of what they saw. All civilisations have invested the heavens with transcendent powers, so that the sky was not merely a physical place, but the home of the gods. Ancient scientists and philosophers even claimed to discover in the stars laws governing the human mind and body, and humanity's origin and future. For many thousands of years, in many different cultures, divination in the form of astrology was an essential part of astronomy. The movements of the heavenly bodies were regarded as a kind of code which the scientist or priest endeavoured to interpret. Hence astronomy was intimately related to religion, both pagan and Christian, and the study of the mechanism of the universe became also a search for the ruler of it.

Celestial maps as we know them emerged during the Renaissance and they would flourish throughout the age of science, but the astronomers of the ancient world undoubtedly possessed various kinds of celestial maps which have survived in fragmentary form. The modern star chart was a product of the Renaissance sense of ordered space, the sense which also saw the development of perspective, terrestrial mapping and scientific diagrams. As they sought ever-greater precision and fullness, the mapmakers of the age of science became engaged in a process of demystification of the heavens. At the same time these maps became a minor art-form, and were overlaid with elaborate constellation imagery, both classical and contemporary. In the nineteenth century, astro-photography began to revolutionise celestial mapping, and deep-space photography now provides much of the fundamental data of modern cosmology. These images from the edge of infinity have had the effect of awakening a new mysticism within the austerities of secular science.

This is not a technical book but I hope that astronomers may still read it with some pleasure. Of all the sciences, the history of astronomy is the most resonant with a sense of mystery and intellectual excitement: I believe that maps and other images of the heavens succeed in some degree in conveying that resonance.

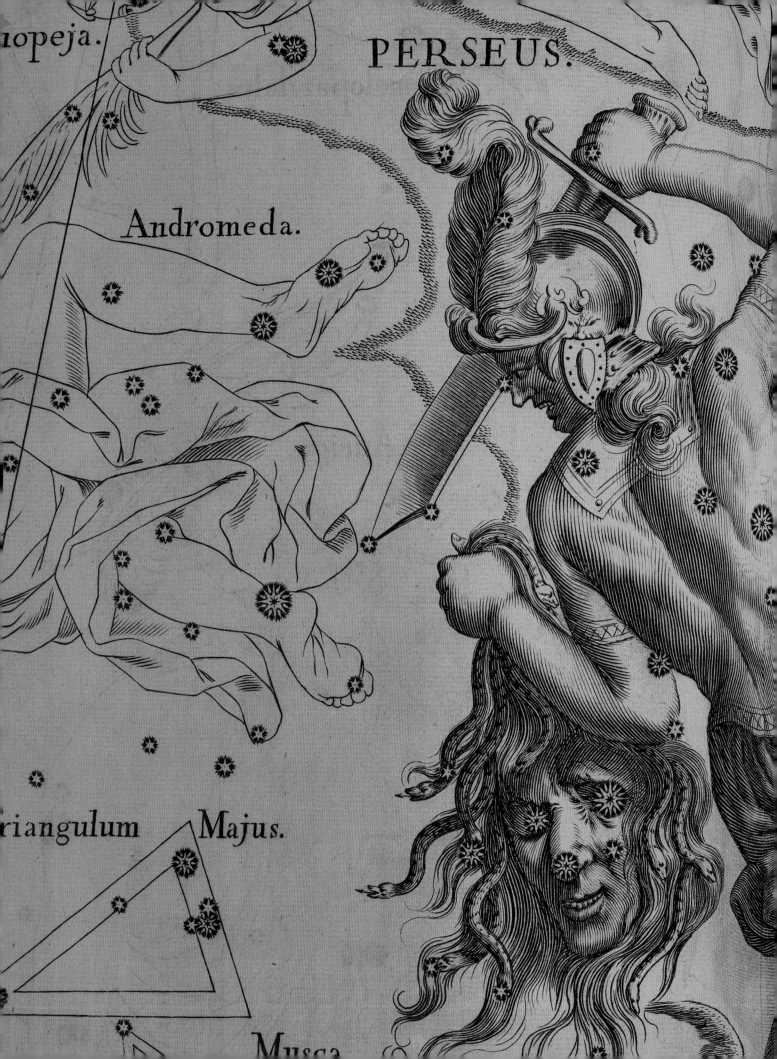

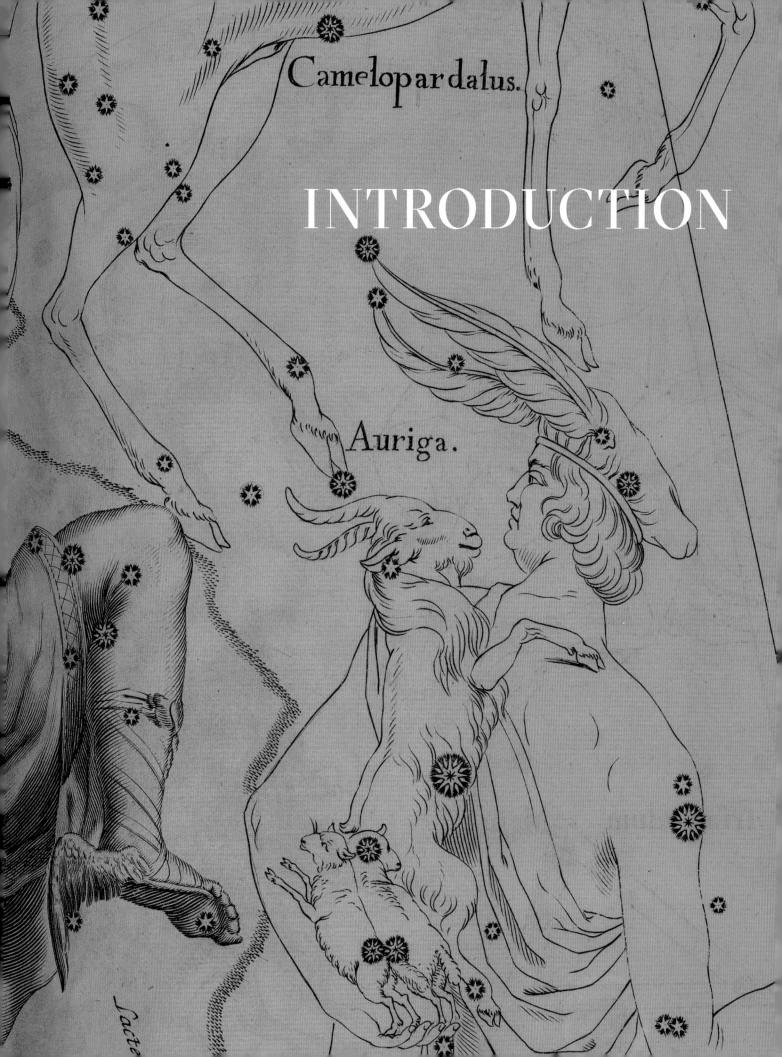

Camelopardalus.

INTRODUCTION

Auriga.

Star Charts and Classical Astronomy

In the pages that follow, some of the images with which astronomers and artists have represented the pattern of the stars in the sky will be described and illustrated. But the way people have seen the heavenly sphere is inseparable from the way they have thought about it, and the subject of heaven is undeniably a large one. The material of astronomy – the heavens – has always been seen to have a spiritual dimension as well as a physical reality. Here we are dealing with evolving science, but also with its interpretation, its images (and these visual interpretations can rarely be divorced from the way artists have visualised other elements of their world), and systems of philosophy and religious belief. This means that a history of star maps, while it must be rooted in the history of astronomy, must also be aware of many wider forces at work in the development of human thought, in particular the fact that its sister-science astrology was for centuries an intricate quasi-religion that was found in many different civilisations.

By contrast, the core period in the history of published celestial mapping – from 1500 to 1800 – coincides exactly with the mature period of terrestrial mapping, and like the map of the world, maps of the heavens acquired a clear identity of their own. They were scientific reference documents, but decorated in the taste of their time; they were published, copied, refined and elaborated as a minor art form. The mapping of the Earth in that period saw the rapid emergence of a consciously scientific approach to cartography through the use of co-ordinates, projections and scales, and progressed to the first exact national topographic surveys. The celestial maps of this period share that cartographic language, handling the techniques of projection, co-ordinates and symbols (though not scale of course) in order to translate information and concepts into graphic form. There is an interesting circularity in this process: techniques such as co-ordinates and projections had formed an important part of mature Greek science, and they had first been developed with reference to the celestial sphere. After a long period of eclipse during the Middle Ages, the rediscovery in the Renaissance of Ptolemy's geography led to the transfer of those techniques from the mapping of the world back to that of the heavens.

Yet of course there were outstanding differences between maps of the heavens and those of the Earth. The terrestrial mapmaker was using mathematical skills, theoretical and practical, to represent the world, or part of the world, in a way that he could never see: his task was conceptual rather than pictorial. No man could see England or Italy or Africa, yet the mapmaker had to draw them, and also to represent on them a wide range of topographic features. The goal of the celestial mapmaker was in a way less

challenging: to record the relative positions of the stars as he saw them, for there is no topography in the sky requiring symbolic representation.

The most ancient images of star groups, for example a Babylonian hunting scene showing the seven stars of the Pleiades in the sky above human figures, began as no more than pictures. Moving forward over the centuries, early medieval manuscripts showing the constellations are still fairly crude pictures. But such simple images are a vast distance away from the mature star chart as it had evolved by the seventeenth century. The essential thing to grasp about the mature celestial map is that it is *not* a picture of what is seen in the sky: it is a conceptual model deliberately plotted in order to display the entire heavens, although usually divided into a northern and a southern sphere. The point of view is an abstract, imaginary point above the north or south celestial pole, looking down upon the starry sphere. The stars are consciously spread outwards from the pole in a projection, which transfers the spherical surface to two dimensions. This model was derived from Greek geometry, and it was used in Islamic celestial globes and astrolabes, which during the Middle Ages were considerably in advance of anything known in the Christian west. Of course the actual material of the star chart – several hundred points to represent the stars –

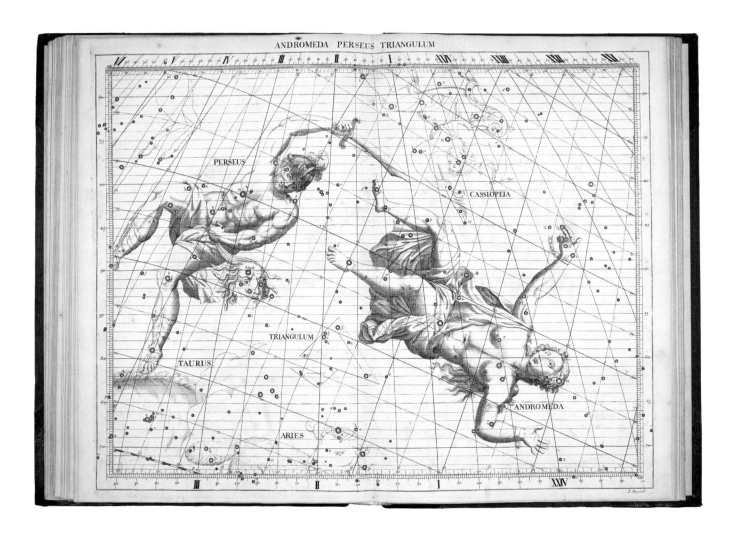

presented no great difficulty to the mapmaker, and was incapable of any great development. But as a conceptual model requiring a mathematical basis and a visual language, the mature star chart of the sixteenth and seventeenth centuries shows a direct application of cartographic technique. There was another type of celestial 'map' which has been arguably even more important historically than the star chart, namely the cosmic model or diagram, in which the relationships between the Earth, Sun, planets and stars are visualised. This form of conceptual modelling cannot be said to be cartographic in the strict sense, but the understanding of these relationships has always been fundamental to astronomy and to human understanding of the cosmos.

There is, however, a more obvious level on which the star chart parallels terrestrial mapping, namely its use of graphic forms and artistic motifs drawn from contemporary models. The hallmark of the star chart from 1500–1800 is the procession of constellation figures, which invariably attracts the eye and dominates the map. Terrestrial maps of this period are likewise instantly recognisable for their engravings of gods and goddesses, ships and sea-serpents, kings and allegorical figures. Yet no group of maps is as exclusively identified with any image as the star chart is with the constellation-figures. The classical constellations were the creations of the cultures of the ancient Near East. The exact origin of most of them is purely conjectural, although it seems probable that they were connected with myths of creation, fertility and the seasons. The origin of even the Greek constellations such as Perseus, whose legends are found in classical writers, cannot be fixed in time, and have parallels in other cultures. The source of the constellation myths has greatly exercised many writers, but they are of no strictly astronomical importance. The fact that the constellations were adopted across cultures – the Babylonian Zodiac reappearing in Egypt and Greece, for example – argues that their cultic or mythic element was secondary. Their great purpose was mnemonic: to define patterns and 'landmarks' in the sky, with which to measure time and direction. The precise resemblance of a group of a dozen stars to a Bear, a Lion or a Hunter was unimportant; it was the pattern that was vital, and the pattern could best be fixed by imposing on the stars certain well-known cultic images. Adopted by the Egyptians, the Greeks and the Arabs, these mnemonic images survived and flourished again in Renaissance Europe as essential features in the structure of the star chart.

It was astrology, of course, which played the major role in reinforcing these images, and centuries of such use made them permanent. But they became in time totally detached from any scientific or cultic basis, and served as purely conventional means to locate any region of the sky. For the astronomer, stars were catalogued as the first, second or third star in Scorpio or Leo and so on, in the manner made standard by Ptolemy of Alexandria (see pages 56–57), while the artistic elaboration of the figures proceeded under its own impetus. The star catalogue listing some 1,000 stars visible to the naked eye, and placed in one of the forty-eight classical constellations, goes back to the forerunners of Ptolemy, and it would remain the fundamental tool of astronomers for some 1500 years, until it was revised

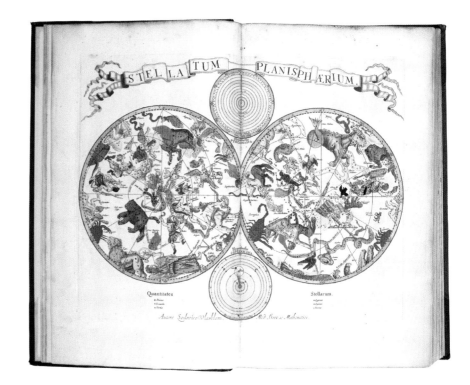

and greatly enlarged in the age of science. The printed celestial maps of the seventeenth and eighteenth centuries were produced largely for an educated but non-scientific market, who saw them as part of their civilised library, along with their world maps, their architectural engravings, their botanical folios and their Latin classics.

This classical connection is vitally significant. Long after the decline of astrology, the constellations were seen by publishers and readers as a series of classical motifs that could be reproduced and elaborated at will. Just as the architect designed classical facades, the engraver illustrated Plutarch's *Lives*, and the poet laboured over his translation of Virgil, so the map publisher perpetuated the forms of the constellations as a conscious link with the classical past. It is almost certain that these twin-hemisphere charts, north and south, were not used by practising astronomers, except as models to demonstrate the general arrangement of the stars; as observational aids they were of no more use than a map of the world would be to a mariner navigating in the Mediterranean. The serious astronomer would use the star catalogue from which the map had been drawn. The great celestial atlases were another matter; the works of Bayer, Hevelius, Flamsteed and Bode were each based on new sky surveys and star catalogues which in their time set new standards of accuracy and fullness. These atlases were composed of a series of detailed charts centred on each constellation, and they may be regarded as the equivalent of the detailed regional maps in a terrestrial atlas.

Both before and after this core period, the pictorial chart of the heavens did not exist, although many other forms of astronomical image did. In the earlier phase the dominant theme is the religious framework within which astronomy and all science was conducted. In the modern age a new

Ursa major

Circulus

CA

TELESCOPIUM
HERSCHELII

LYNX

AUR

Apollo Castor

Lux Abrachaleus

Cuetos Sum   Cassiopeja   Lacte

OPARDALUS

Arcticus

Perseus

Algenib   Caput

Algol
variabilis

Menkib

Atik

Vit

Plejades

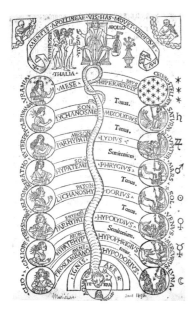

generation of photographic star maps and images has appeared which have
reached out beyond the visible sky and formed the basis of a new cosmology.
This book concentrates on European astronomy, which has developed
continuously down to the present day. In the past, non-European traditions
achieved levels of skill in astronomy that were in advance of European
practice, but they invariably reached a point where they ceased to develop;
there was no Copernican revolution or scientific revolution outside Europe.

In one sense astronomy is science at its most austere: the desire to
observe, chart and understand a realm of the universe that was permanently
beyond the reach of direct experience was driven by intellectual curiosity
in its purest form. Yet there was a practical and universal motive behind
the early study of the heavens: namely, to measure time. All civilisations
perceived the cyclical pattern in the night sky, and seized upon it as a means
of measuring and anticipating the passage of the hours, days, seasons and
years. And beneath this technical function lay the religious dimension which
all cultures saw in astronomy, and which gave a spiritual urgency to the
astronomical quest. The heavens are not merely a physical place; they form
a divine realm where supernatural powers dwell and reign over the Earth.
The study of the mechanism of the universe became also the search for the
creator and sustainer of it. In one of the most famous metaphors in classical
literature, Plato compares man's limited perception of reality to that of
prisoners chained in a cave:

> The visible realm corresponds to the prison, and the light of the fire in the
> prison to the power of the Sun ... the ascent into the upper world and the
> sight of the objects there, represent the upward progress of the mind into
> the intelligible realm.
> Plato, *Republic*, Book 7

Plato's fable of the cave is an apt metaphor for man's early attempts to
understand what he saw in the sky. Although not so confined in vision as
the cave-prisoners, the ancient observer witnessed an apparent procession
of shapes and movements in the heavens, whose true interpretation defied
him. The crucial fact – that the apparent movements of the heavenly bodies
are largely imparted by man's shifting viewpoint – was not grasped until the
sixteenth century. Cosmic models before Copernicus naturally assumed an
unmoving Earth at the centre of the universe.

The number of stars 'visible to the naked eye' is of course debatable:
the maximum figure is perhaps 15,000 (in the whole sky, not all visible
together). But even a practised observer can readily identify only a few
dozen individual bright stars, and all others must be found by reference first
to their brighter neighbours and to the star groups within which they lie.
Hence the designation of constellations was a referencing system which all
cultures have found essential. The greatest star catalogue of antiquity, that of
Claudius Ptolemy c. AD 150, located slightly more than 1,000 individual stars,
and this remained the canonical number until the eighteenth century, when
the southern stars were charted and when Flamsteed catalogued some 3,000

stars of the northern sky. Until the invention of the telescope in the early seventeenth century enabled fainter stars and other objects to be seen, the material of astronomy – the face of the heavens – remained what it had been for some 4000 years of astronomical history.

The Starry Sphere

To any observer, ancient or modern, the sky appears as a dome resting on the surrounding horizon. In the course of each night the stars, fixed in an unchanging pattern in relation to each other, appear to move *en masse* across the sky. The movement is not linear, but appears to rotate about a fixed point above one of the Earth's poles. To an observer in the northern hemisphere, part of the sky around the north pole is visible throughout this rotation, another section around the south pole is always invisible, while the rest of the sky appears to rise and set each night. The size of these ever-visible and ever-invisible spheres will vary with changing latitude; only on the equator will the entire sky be displayed. Ancient cultures conceived the stars to be set in a starry sphere, and it was this sphere which was thought to revolve around the Earth, although what that sphere was, its physical reality, was always a mystery. The concept of the starry sphere is still a highly practical one for the observer and the mapmaker.

In addition to this daily revolution, the starry sphere was apparently also shifting in a second and much slower cycle. Stars visible at midnight moved one degree westward by each succeeding midnight. New stars appeared on the eastern horizon each nightfall, while in the west other stars vanished and were no longer seen after sunset. In 180 days the rotating sphere of stars changed its aspect completely. Astronomers in some cultures appreciated that this cycle was related to the Sun's movement: that the stars were still present in the daytime, but were lost in the stronger light of the Sun. Stars visible at midnight will be hidden by the noonday Sun 180 days later, so that, for example, the Sun is said to be 'in Aries' when the stars of Aries are invisible in May. We would now express this as a function of the Earth's revolution around the Sun, but the effect is the same.

An older and quite universal discovery was that five stars were not fixed like the others, but moved independently across the starry sphere. These planets (from the Greek word meaning wanderer) moved in a most puzzling way, each appearing for some period of time to slow down and then reverse its direction through the stars before moving forward once more. The Sun-centred orbits of the five planets – Mercury, Venus, Mars, Jupiter and Saturn (Uranus, Neptune and Pluto were undiscovered) now serve to explain these motions, as the planets overtake or are overtaken by the Earth. But this planetary behaviour caused the greatest difficulties to all early astronomical theorists. Like the stars, the Sun, Moon and planets appeared to revolve independently about the Earth in their daily motion, but in addition, over the course of a year, their paths appeared to rise and fall, within certain limits, against the background of the fixed stars. To explain this, the spheres bearing the Moon and planets were conceived by Hellenic theorists to bear secondary cycles – called epicycles – to account for this dual motion. The fixing of their paths against the starry sphere, the precise modelling of the supposed epicycles, became one of the principal tasks of classical astronomy. It must be emphasised that the charting of the yearly cycle of the stars, and of the more complex paths of the planets, was not dependent on the modern knowledge that the Earth revolves around the Sun: what we see is still the same.

The Sun's apparent path through the stars in the course of the year was an important discovery, but it is perhaps not immediately obvious why, since the Sun and stars are never visible together. The answer is that this path, called the ecliptic, is a line that is fixed in the sky, and it became the baseline for a locational framework in the heavens. The horizon is of course always shifting, but the ecliptic is an objective reality, which in the context of Greek geometry became the basis for a mature referencing system. The extremes of the ecliptic, the highest and lowest points touched by the Sun, were called the tropics: they coincided closely with the region of the sky within which the Moon and planets reached their extreme positions, and this part of the sky attracted the greatest attention. As the geometry of eclipses came to be understood, it was seen that they occurred only in this zone, hence its name. When moonlight was recognised to be reflected sunlight, it was understood that the full Moon was directly opposite the Sun, while the new Moon was hidden against the Sun. We now know that the Sun's apparent rise and fall along the ecliptic is caused by the Earth itself, whose axis is tilted at roughly 23 degrees from the perpendicular. As the Earth slowly circles the Sun, the observer will see the Sun apparently rise and fall and rise again in an annual cycle. Age-old myths such as that of Persephone and Hades expressed an intuitive perception of the cyclic pattern in time and nature; but well before the fifth century BC it was understood, on a scientific level, that this rise and fall of the Sun was the true cause of the seasons. We could express this by saying that if the Earth's axis were not tilted at 23 degrees to the vertical, there would be no seasonal changes on the Earth, for this tilt brings the regions of the Earth's surface into the Sun's heat then out again in the course of the year.

One further type of celestial motion was observed by the Greeks and perhaps by others too, namely that the whole starry sphere is slowly shifting westwards. The stars drift in paths parallel to the ecliptic by about one degree in seventy-two years. The movement, now known as the precession of the equinoxes, has important consequences for the later mapping of the heavens and it will be referred to many times in this book. Its cause is a slight perturbation of the Earth's axis: the planet's poles are revolving slowly in a circle with a radius of 23 degrees, in the way that the apex of a spinning top revolves. A precessional cycle, carrying the stars round the sky completely, would take almost 26,000 years. This movement renders an accurately drawn star chart valid only for a limited period of time, say 50–100 years. Its most noticeable effect is that the pole star is only the pole star for a given epoch, until the north pole moves gradually into a new alignment. The present pole star, Polaris, will be most directly above the north pole in the year AD

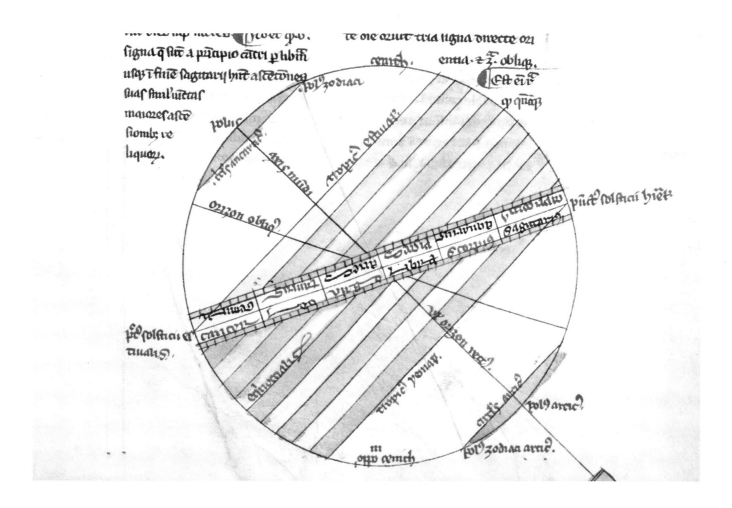

Above

The Earth's sphere mapped to show its relation to the celestial sphere: the circles of the equator, the tropics, the polar zones and the ecliptic. This manuscript of Sacrobosco's fundamental textbook of astronomy was produced in Paris in the late 13th century; it incidentally nails the myth that people in the Middle Ages believed the Earth to be flat.

Overleaf

Plate from Cellarius's *Atlas Coelestis* (1660).

2100. In the Old Kingdom of Egypt (2686–2181 BC) the star Thuban in the constellation Draco was the pole star, and some pyramids appear to have been aligned with its historic position at that time.

In the evolution of genuine celestial maps, a reference co-ordinate system was essential. The simplest would be a horizon system where positions are given in altitude degrees above the horizon plane, and in azimuth (direction) clockwise around the horizon starting from an agreed, predetermined point. But it is clear that such a chart would be valid only for a given latitude, a certain season and indeed a particular time of night. It would be essentially a diagram of one part of the sky at one moment only; a whole series of such maps would be required to cover a year, and even then they could be used only in their place of origin. These problems were to be ingeniously addressed in the design of the astrolabe (see page 36–37). A more comprehensive and objective system of star mapping is obviously desirable. The significance of the ecliptic is that it provided a baseline in an objective framework upon which the first maps or globes of the sky might be drawn.

Throughout late classical, medieval, and early modern astronomy, celestial latitude was measured north or south of the ecliptic plane. Celestial longitude was measured east from the point where the ecliptic crosses the plane of the equator, known as the first point of Aries. This marked one of the great turning points of the year, the spring equinox, which now takes place on or

Plate II.

Opposite
Drawing of a typical celestial globe,
produced for scholars, scientists and
navigators, from Benjamin Martin's
*The Description and Use of Both the
Globes, the Armillary Sphere
and Orrery, c.* 1760.

near 21 March. At the time when the astrological star signs of the Zodiac originated, around 300 BC, they coincided with the relevant constellations. But by the precessional movement described above, these constellations have now shifted some 30 degrees westward. However, the traditional signs of astrology have not been moved, so that the 'sign of Aries' is now occupied by Pisces. The spring equinox will be contained in Pisces until *c.* AD 2700 when it will move into Aquarius. The following table illustrates this important point.

Constellation	Current Solar Passage	Containment of Vernal Equinox	Astrological Sign Period
Aries	19 Apr – 14 May	2000 BC – AD 100	21 Mar – 19 Apr
Taurus	15 May – 20 June	4500 BC – 2000 BC	20 Apr – 20 May
Gemini	21 June – 20 July	6600 BC – 4500 BC	21 May – 21 June
Cancer	21 July – 10 Aug	8100 BC – 6600 BC	22 June – 22 July
Leo	11 Aug – 16 Sept	10800 BC – 8100 BC	23 July – 22 Aug
Virgo	17 Sept – 30 Oct	AD 12000 – AD 15300	23 Aug – 22 Sept
Libra	1 Nov – 23 Nov	AD 10300 – AD 12000	23 Sept – 23 Oct
Scorpio	24 Nov – Dec 17	AD 8600 – AD 10300	24 Oct – 21 Nov
Sagitarius	18 Dec – 19 Jan	6300 AD – AD 8600	22 Nov – 21 Dec
Capricorn	20 Jan – 15 Feb	4400 AD – AD 6300	22 Dec – 19 Jan
Aquarius	16 Feb – 11 Mar	AD 2700 – AD 4400	20 Jan – Feb 18
Pisces	12 Mar – 11 Apr	100 BC – AD 2700	19 Feb – 20 Mar

The other fundamental requisite for celestial mapping was some form of zoning or landmarking by which the observer could orientate himself, to compensate for the absence of topography in the sky. The origin of constellations is a problem that belongs perhaps to the study of mythology and anthropology rather than astronomy; but the eighty-eight star-groups now accepted by modern astronomers fall into fairly distinct historical groups, as the table on pages 26–28 shows.

The 88 Constellations with some Principal Stars

The Ptolemaic Constellations: The Zodiac

Aries	Ram	
Taurus	Bull	Aldebaran, Pleiades
Gemini	Twins	Castor, Pollux
Cancer	Crab	
Leo	Lion	Regulus
Virgo	Virgin	Spica
Libra	Scales	
Scorpius	Scorpion	Antares
Sagittarius	Archer	
Capricorn	Sea-Goat	
Aquarius	Water-Bearer	
Pisces	Fishes	

Northern Ptolemaic Constellations

Andromeda	Andromeda	M31 Galaxy
Aquila	Eagle	Altair
Auriga	Charioteer	Capella
Bootes	Herdsman	Arcturus
Cassiopeia	Cassiopeia	
Cepheus	Cepheus	
Corona Borealis	Northern Crown	
Cygnus	Swan	Deneb
Delphinus	Dolphin	
Draco	Dragon	Thuban
Equuleus	Little Horse	
Hercules	Hercules	
Lyra	Lyre (-Bird)	Vega
Ophiuchus	Serpent-Handler	
Pegasus	Winged Horse	
Perseus	Perseus	Algol
Sagitta	Arrow	
Serpens	Serpent	
Triangulum	Triangle	
Ursa Major	Great Bear	
Ursa Minor	Little Bear	Polaris

Southern Ptolemaic Constellations

Ara	Altar	
Argo Navis	Ship	(Now divided Into four)
Canis Major	Great Dog	Sirius
Canis Minor	Little Dog	Procyon
Centaurus	Centaur	Alpha
Cetus	Whale	
Corona Austrina	Southern Crown	
Corvus	Crow	
Crater	Cup	
Eridanus	River	Achernar
Hydra	Water Snake	
Lepus	Hare	
Lupus	Wolf	
Orion	Hunter	Rigel, Betelgeuse
Pisces Austrinus	Southern Fish	Fomalhaut

Southern Constellations Added *c.* 1600

Apus	Bird of Paradise	
Chameleon	Chameleon	
Dorado	Swordfish	Large Magellanic Cloud
Grus	Crane	
Hydrus	Water-Snake	
Indus	Indian	
Musca	Fly	
Pavo	Peacock	
Phoenix	Phoenix	
Triangulum Australe	Southern Triangle	Small Magellanic Cloud
Tucana	Toucan	
Volans	Flying Fish	

Constellations of Jakob Bartsch, 1624

Camelopardalis	Giraffe
Columba	Dove
Monoceros	Unicorn

The 88 Constellations with some Principal Stars (continued)

Constellations of Hevelius, 1687

Canes Venatici	Hunting Dogs
Lacerta	Lizard
Leo Minor	Small Lion
Lynx	Lynx
Scutum	Shield
Sextans	Sextant
Vulpecula	Fox

Ancient Star Groups Now Reformed

Carina	Keel of Ship	Canopus
Coma Berenices	Berenice's Hair	
Crux	Southern Cross	
Puppis	Stern of Ship	
Pyxis	Compass of Ship	
Vela	Sail of Ship	

Southern Constellations of Lacaille, *c.* 1750

Antlia	Pump
Caelum	Chisel
Circinus	Compasses
Fornax	Furnace
Horologium	Clock
Mensa	Table
Microscopium	Microscope
Norma	Square
Octans	Octant
Pictor	Easel
Reticulum	Reticle
Sculptor	Sculptor's Workshop
Telescopium	Telescope

The Polar Stereographic Projection

Printed star charts of the northern or southern hemispheres shared the fundamental structure of the astrolabe rete (see page 71). Neither is a picture of what the human eye sees in the heavens, but is a sophisticated geometric structure, modelled in order to transfer the sphere of heaven on to a two-dimensional plane. The principle is the polar stereographic projection, in which the stars are projected on to the plane of the equator. The point of origin of the projection is the north or south celestial pole, from which lines are taken to a number of key points on the celestial sphere – the tropics, equator and ecliptic. Points are marked where these lines intersect with the plane of the equator. That plane, viewed from the perpendicular, then becomes a two-dimensional projection of the celestial sphere. The position of all the required stars can be plotted using this method. The result is an ingenious spreading of an entire hemisphere, greatly extending what the Earthbound observer can see, and extending also what an observer above the pole of a celestial globe will see. The farther a star lies from the pole, the farther away it will appear on the plane of projection. In theory the entire heavens might be shown, although the spreading effect would become enormous towards the south pole; indeed, the pole itself would become not a point but a circle, forming the boundary of the map. In practice the projection was extended only as far as the Tropic of Capricorn, which formed the edge of the astrolabe. Many printed star maps, such as Albrecht Dürer's, show the ecliptic as the map's border, in which case the point of projection must be the ecliptic pole. On this projection, the ecliptic becomes an eccentric circle touching both tropics at the solstices, and this is an essential feature of the astrolabe. The theory of the stereographic projection was known to Hipparchus of Rhodes around 150 BC, and thereafter to Ptolemy. The earliest surviving post-classical account of it was written in Alexandria in the sixth century AD. In the form of the astrolabe, the projection was in continuous use by Islamic scholars from the eighth century. In the west, Latin treatises on the astrolabe described it from the eleventh century onwards. This use of geometric principles to create a form of co-ordinate mapping, familiar throughout the Middle Ages, forms a striking contrast to terrestrial mapping, where such methods were unknown, in western or in Islamic science. It is an open question why celestial maps drawn in this way have not been discovered in any manuscripts, western or oriental, earlier than about 1440, and why it was used only to make the astrolabe rete. When two star charts, one of the northern heavens and one of the southern, were placed side by side, the result was a double-hemisphere map of the entire heavens, paralleling the familiar double-hemisphere maps of the Earth.

Above
The astrolabe explained in a medieval Latin translation of the work of Masha'Allah, an Arab astronomer of the eighth and early ninth century; this was the text used by Chaucer in the first English treatise on the astrolabe, which he composed in the 1390s.

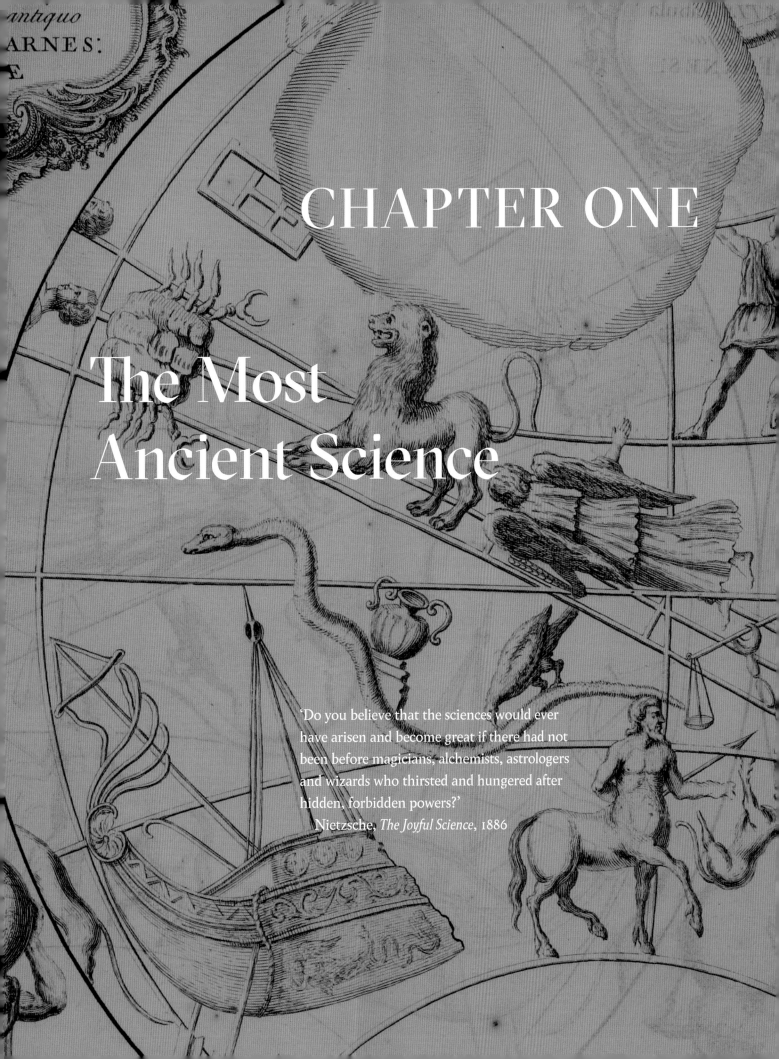

CHAPTER ONE

The Most Ancient Science

'Do you believe that the sciences would ever have arisen and become great if there had not been before magicians, alchemists, astrologers and wizards who thirsted and hungered after hidden, forbidden powers?'
Nietzsche, *The Joyful Science*, 1886

Order and Mystery

Odysseus
Gazing with fixed eye on the Pleiades,
Boötes setting late, and the Great Bear ...
Looking ever towards Orion ...
Homer, *Odyssey*, Book V

The homeric poems are the oldest creations of European literature, dating from not later than 800 BC, yet the astronomical knowledge embodied in these lines is at least 1,000 years older still. Astronomy is by far the oldest exact science, fulfilling perfectly the simplest definition of science as knowledge of the natural world, its regularities and patterns. All ancient cultures observed the stars and recognised their cyclical changes. Through patient, naked-eye observation over centuries, practised star-watchers in Europe, Egypt, Mesopotamia, China and Central America built up a knowledge of the sky that was in some ways more detailed than their knowledge of the world in which they lived. In ancient societies, indeed in all pre-industrial societies, real darkness still filled much of people's lives, and the farmer, priest or seafarer was familiar with the star-patterns as the fundamental gauges of time and direction. All cultures have identified star groupings, constellations, in the sky. Their resemblance to their supposed subjects was often tenuous, but they were essentially mnemonic devices serving to unify groups of stars, and impose some order on the ocean of the night sky.

Yet alongside this precise, observed regularity was another very different element: the sense of the mystery of the sky, a feeling that it was a realm quite other than this Earth, a realm where divinities and spirits had their home, for the most part dwelling in icy serenity, but not infrequently unleashing storms, and exercising a mysterious power to shape human lives. It is this duality of precise observation and religious awe which gives the early history of astronomy a double fascination. In the ancient Near East, where recorded astronomy begins, the agency of gods or demons was seen in every aspect of life and nature. Mesopotamian civilisation perceived in the sky one of the most fundamental patterns of nature, for the sky was the great time-measurer: the alternation of day and night, Sun and Moon, summer and winter, divided and regulated the endless flow of time. All this was recorded and mastered in detail. Yet beneath these observed patterns there has always been the search for causes. The causes advanced by thinkers throughout much of human history would now be rejected as illusory: the overwhelming cause assigned by pre-critical science was the will of the gods. But even the

gods did not offer a universal escape from the exercise of reason, for the question inevitably arose whether the gods and spirits wielded arbitrary, irrational power, or whether they too obeyed fundamental rules.

All ancient cultures, except perhaps the Chinese, regarded the heavenly bodies as divine, and their regularity and serenity, so different from the striving and suffering of life on Earth, seemed to hint at a fundamental order lying at the heart of nature. This co-existence of science and religion can be seen in many creation myths, in which the crucial element is the imposition of order on primal chaos. In the Babylonian creation epic *Enuma Elish* (c. 1800–1500 BC), Marduk kills the unruly goddess of primal waters, Thiamat, and cuts her body in two, using one half to create the heavens and the other the Earth. The constellations and the movement of Sun and Moon were to be under Marduk's beneficent rule, and to mark the passage of time. Such myths reveal a sense of, or perhaps search for, order in the universe, amid numerous elements of chaos. This same order-seeking impulse lay behind the cosmic diagrams drawn in many ancient cultures, which depicted some of the fundamental elements of Earth and sky, and sought to locate humanity in space and time.

The very earliest evidence of astronomical awareness is just as old but is more difficult to interpret. Many of the megalithic stone structures of northern Europe are unmistakably aligned with heavenly bodies, and with a precision that argues for a mature tradition of observation. Stonehenge's alignment with the midsummer sunrise is the most famous, and a well-developed calendar, and perhaps also some form of Sun-cult, must surely have existed among the people who built and used Stonehenge. These stone or earth structures were built in order to create an artificial horizon against which astronomical events could be precisely observed; they were in a sense instruments. Some of these 'instruments' are aligned on objects more subtle and surprising than the Sun: in north-east Scotland there are stone-groups in a recurrent pattern of two erect columns framing a horizontal, recumbent stone, which makes a clear, level horizon. Some of these stones appear to register the upper and lower limits of the Moon's orbital plane, a movement which occupies an eighteen-year cycle, and it seems impossible that these alignments could be accidental. These stones probably date from around 2000 BC onwards, and if they have been correctly interpreted, they argue a degree of sophistication in their builders' knowledge of the skies that is truly startling, and one which must have been built up over centuries of painstaking observation. There are many still more enigmatic examples too. The so-called White Horse at Uffington in Berkshire, a figure cut into the chalk hillside, may not be a horse at all, since it appears to be connected with the constellation Taurus, and with the rising of the bright star Aldebaran – 'the eye of Taurus' – over the animal figure. In the absence of archaeological or written records, it is impossible to interpret such sites or to infer the true motives of the ancient astronomers. The most likely explanation of such alignments is that they were gauges of time: when the Moon or Sun or a certain star reached a predetermined position, it was known that the year had elapsed and that the cycle would begin again. But how and why were the complex cycles of more than a year studied and interpreted? They clearly might form part of a belief system where the order which the observers found in the heavens was indicative of something fundamental, if mysterious, in the universe. The roots of astronomy have always been both practical and religious: it was studied and used, and it fed the mind and the imagination.

Measuring Time

The crucial practical use to which astronomy was put, and which provides evidence of observational science in virtually all ancient cultures, was the making of calendars. It was widely noticed that the basic time-units of the day and the year did not precisely divide, the one into the other, and moreover that the intermediate unit, the lunar month, did not match with the solar cycle. All the complexities of the calendar result from these irreconcilable time periods. All cultures have sought to find greater time periods or cycles in which these three would coincide, before diverging again. Such a cycle could be used to tie days, years, dynasties, eclipses and religious ceremonies into a greater historical framework. Some of the ancient cycles occupied centuries or even millennia, and their calculation involved observing the stars and planets over many generations, and required powerful mathematical systems to extend the results far into the future. For example, the Egyptian civil calendar early in the third millennium BC adopted the round figure of 365 days for a year. But since the natural year is actually closer to 365.25 days the civil calendar began to creep slowly forward through the natural year. The Egyptians noticed this and calculated correctly that in 1460 years the calendar would coincide with the true sidereal year, the year of the star-cycle. This was known as the Sothic cycle, after the bright star Sothis, known to us as Sirius, which played a vital part in their year.

But however sophisticated their calendars, these ancient sciences have left almost no evidence of their work in graphic form: there are no genuine star charts extant from ancient Mesopotamia, Egypt, India, China, or Mexico. Pictures and diagrams, some childishly simple, others more complex, are all that have survived. These ancient societies had not developed cartographic awareness or a cartographic language in general, in relation to terrestrial mapping, so that it would be unreasonable to expect the concept of mapping to be applied to the heavens. In all pre-Hellenic sciences, it seems that the key which would underlie accurate celestial mapping – that of spherical geometry – was absent. The mathematical systems which these cultures built up were highly advanced, but if they proposed models of the cosmos or diagrams of parts of the heavens, they were imaginative or symbolic; to the priestly elite who studied the skies, diagrams of the star positions seem to have been unknown.

Another important function of ancient star-watching, but one in which even less specific evidence has survived, is navigation. Literary references in Homer show clearly that Mediterranean sailors used the Sun and stars, especially the pole star, for direction-finding, but their skills were

Right & Next Spread
Egyptian constellations from the tomb of Seti 1 (1294–1279 BC). These northern constellations appear in various contexts in Egyptian art. The only certain identification that can be made is the Ox and his Handler, where our Plough (Ursa Major) can be clearly seen. Unfamiliar shapes such as the crocodile, the falcon-headed god Horus and the goddess Taweret, in the form of a hippopotamus with a crocodile on her back, do not correspond to any classical constellation. These star-groups would later be integrated with the Babylonian Zodiac figures.

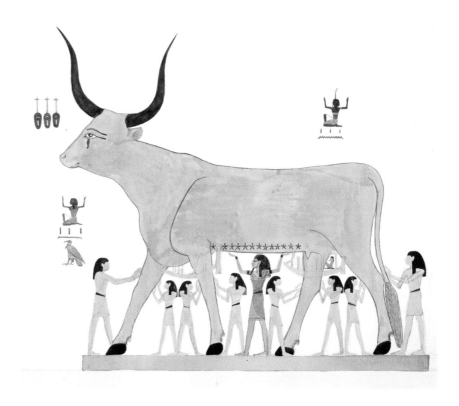

undoubtedly handed down through oral tradition rather than documents or maps. When seafaring lore was first written down, the earliest graphic form to emerge was not the star-map but the wind-rose, a type of compass whose points were the four or eight winds, for the skilled mariner could readily distinguish between the cold northerly wind and the warm southerly, and set his course accordingly.

The Foundations of Astronomy: Egypt and Mesopotamia

Despite this absence of graphic images which we would recognise as star maps, there is ample evidence of astronomical science contemporary with the Neolithic structures of northern Europe, and with the invention of writing in Mesopotamia and Egypt in the third millennium BC, it becomes possible to interpret both knowledge and beliefs. In Egypt as early as the Old Kingdom (c. 2686–2181 BC) a form of stellar reference system had been devised, to which the term 'decans' was later applied. The decans were thirty-six star groups in the vicinity of which the Sun rose in the course of a year. They functioned effectively as a constellation system, as landmarks in the sky, and their appearance as decorations on temple ceilings, tombs and coffin lids may be considered the oldest astronomical pictures in the world,

Below
The planets as Egyptian animal-headed gods, shown riding in boats through the night, from the walls of the Temple of Dendera.

although whether they can be regarded as genuine maps is a matter of debate. The decans became associated with various deities and are represented as individual figures, as are the planets. The decans were all in a band just south of the ecliptic, but other major northern constellations became fixed too and recur in recognisable form in paintings spread over many centuries. The striking thing about them is that, with a very few exceptions, they cannot be identified with the classical constellations with which we are familiar. This tends to upset any belief that the constellations are in any sense objective or inevitable. The planets, however, seem always to have been regarded as personal deities, as they were in most other cultures, and are shown as animal-headed gods upright in the boats in which they journeyed through the night. 'Horus the Red' is clearly Mars, while 'The Crosser' with two faces is Venus, recognised as both the morning and evening star.

The Egyptian civil calendar was built around the decans, which formed thirty-six 'weeks' each of ten days, with five intercalary days added to complete the year. In Egypt the turning points of the solar year, the summer and winter solstices, lacked the importance which they carried for more northerly cultures. Instead the annual flooding of the River Nile in July was the crucial event which irrigated the land, and this was seen to coincide with the first appearance each year of the star Sirius (in Egypt Sothis) after its period of invisibility below the horizon. Thus the festival of Sothis each summer was the year's great turning point, in the month called 'The Opener of the Year'. This first pre-dawn rising of a star after the period during which it rose in daylight is known as heliacal rising, and it was an important concept to the early Egyptians. Each day after its heliacal rising, a star would rise slightly more in advance of the Sun, until another suitably conspicuous star rose heliacally. The decans centred around such stars, and the Egyptian map of the heavens hinged on this concept of star rising, which in an intriguing way also left its mark on the hour divisions of the day and night.

The Egyptians were deeply interested in the passage of the Sun-god Re through the underworld at night, on his ship with his attendant deities before his re-emergence in triumph each dawn. The underworld was considered to be divided into twelve regions in each of which he spent a short period of time. To measure these periods, the nightly rising of twelve stars was fixed upon, and the interval between them was termed an hour, the hieroglyph for which is a star. From this system the Egyptians devised star clocks which appear on many coffin-lids from 2200 BC onwards. They take the form of columns of star names for each hour of the night and for each decanal week of the year, so that by matching the star seen rising in the sky with its symbol in the table, the sky became to the skilled observer literally a clock. It is interesting to note that this type of star clock vanished c. 1500 BC, presumably because those who needed to do so had mastered this symbolic map and could read the hours directly from a glance at the sky. The twelve-hour division of the day was apparently purely by analogy with the night, and shadow clocks were marked with five divisions: the shadow crossed the five markers as it shortened towards noon, then re-crossed them as it lengthened towards evening, with one hour of twilight both for sunrise and for sunset.

Above
The Sun-cult of the king Akhenaten from his new palace at Amarna, c. 1340 BC. The king and queen are seen receiving the divine life-force from the Sun's rays. The new monotheistic religion of the Sun was embodied in a new royal city and temples, built away from the centre of the old religion at Thebes. They were aligned with the Sun, and decorated in a vivid new style.

The cult of the Sun left its mark on the architecture of Egypt as it did on that of northern Europe. The temple of Amun-Re, the Sun-god, at Karnak (modern Luxor) was aligned so that the setting Sun shone through an axis corridor into the sanctuary of the temple. The celebrated apostate king Akhenaten (1352–1336 BC) instituted a new monotheistic worship of the Sun, and his new palace at Amarna was richly decorated with scenes of worship of the Sun-disc, the Aten. The new cult further stimulated astronomical observation, but it was short-lived, dying with the king himself. The astronomical significance of the pyramids themselves is uncertain. For all their evident scientific skills, the Egyptians left no precise records of astronomical observations, no star catalogue, no tables of star risings, planetary movements, eclipses and so on. Nor did they devise anything in the nature of astronomical theory: there was apparently no speculation, scientific or religious, as to the causes of what they observed in the sky. Instead Egyptian astronomy was intensely practical, whether that practicality was in the field of time-measurement or divination.

The true origins of classical astronomy lay elsewhere in the ancient Near East, in Mesopotamia, where detailed observations and calculations were first recorded, and where it becomes possible for us to recover a system of beliefs about the heavens. Some of the mathematical methods of the Babylonians were extremely powerful and sophisticated; the constellations they designated proved permanent, and their philosophical view of man's relation to the cosmos laid the foundations of astrology.

The very earliest explicitly astronomical texts are Babylonian, dating from c. 1500–1700 BC. They are clay tablets written in the cuneiform script, and they take the form of omens in which the positions of Sun, Moon, stars and planets are related to events such as wars, famines, and royal successions. Thus these earliest texts make the fundamental link between the precise science of astronomical observation and the art of divination, the two forces that were to motivate astronomy for the next 3,000 years. So deep was this

The Babylonian Sun god, Shamash, is seen sawing his way up through the eastern mountains, with Ishtar (Venus), goddess of the morning star, before him. Rivers, in the form of Ea, god of fresh water, run down from the mountains. Impression from a cylinder seal, c. 2300 BC.

early interest in divination that one of the strongest motives for the constant observation of the sky was the hope of obtaining warnings of future events in nature and in human life. In this early period the Zodiac does not appear fully; the Moon, planets and some constellations form the material basis of the omens. Star figures such as the Lion, Bull and Scorpion make their earliest appearance here, and they appear in Babylonian sculptures in other contexts too. It is interesting to note that the brightest stars in these groups, together with that of Pegasus, are almost exactly 90 degrees apart on the Zodiac circle, and their heliacal risings coincide with the four turning points of the year: the spring equinox, summer solstice, autumn equinox and winter solstice. This suggests that the framework at least of the Zodiac structure had already been recognised. The omens however are based on visible events current in the sky, not on calculations of invisible influences. The stars were termed 'the gods of the night', while the Sun was depicted as a regal figure rising each dawn from a mountain range. At this early stage the Babylonians, like the Egyptians, seem to have believed that the Sun passed each night in an underworld, where the stars probably stayed during the day.

From this earliest date the Babylonians identified the heavenly bodies as in themselves divinities able to affect human life. Indeed the belief in the *personality* of the heavenly bodies was the essential starting-point for what was later to become astrology. It was the deities themselves, such as Marduk or Ishtar (corresponding to Jupiter and Venus) who might determine matters of politics or love, and their will was thought to be discernible through their celestial behaviour, especially whether they were high and dominant in the sky. The notion that the planets could control human events merely *as planets*, by virtue of their astronomical position alone, is secondary, and it is also far less rational. The prediction of human events by planetary position depended on an exact knowledge of the heavens, but paradoxically the later accumulation of that knowledge served to obscure the original belief in the personality of the celestial deities. Once the personal aspect of heavenly bodies is rejected, the crucial question prompted by these omens, by the belief in celestial divination, is by what power do the stars and planets actually cause events? It is possible to see astrology as a belief system appropriate to a wholly mechanistic universe where man has no free will, and where perhaps the divine powers are also fixed in their roles. This vision would later present grave difficulties for all the mature religions.

In the early Babylonian phase, the omens, as manipulated by the priesthood, functioned mainly to guide the court in matters of politics, war and personal fortune. They seem to have been regarded as indicative only, not as revealing inexorable fate, so that religious observances and magic might avert the omen. The Assyrian king Esarhaddon (680–669 BC) was so fearful of lunar eclipses that during his reign he enthroned substitute king-figures when eclipses occurred, who were afterwards executed to divert the malign influence of the eclipse from the king himself. The Babylonians' skills in astronomy and their astral religion both left their mark on the Old Testament: the prophets' horror of star-worship is expressed by Isaiah: 'Let now the astrologers, the stargazers, the monthly prognosticators, stand up and save thee from these things that shall come upon thee. Behold they shall be as stubble; the fire shall burn them ...'

Despite such deep levels of superstition, Babylonian astronomy was based on rigorous observation and on advanced mathematical skills. Like all ancient civilisations their observations were made only with the naked eye, but they built structures which functioned as artificial horizons with fixed observation points, which may be considered as early instruments. Detailed computations of stellar and planetary motions formed the basis of their calendars, as they did later of the art of astrology. Remarkably, the Babylonians developed no form of spherical geometry or trigonometry: celestial positions and timings were all predicted mathematically, and the concept of star maps or cosmic models was apparently quite unknown. The method used to track the heavenly bodies was to record the times at which they appeared at two or more cardinal points, then to calculate all its further positions as functions of time, in constant progression. Allowance was made for the changing velocities of the planets, which the Babylonians were able to tabulate in almanacs. This method was applied with precision and sophistication, and led to what was probably the most important development in classical astronomy: the designation of the Zodiac. It had already been noticed that the Sun dwelt for three months in each quarter of a celestial circle marked by the cardinal points Scorpio, Leo, Taurus, and (at that time) Pegasus. If each of these sky-regions were further divided into three, corresponding to the

Below, Opposite & Next Spread
Chinese star chart, AD 940. The oldest surviving paper star map from any civilisation. The Great Bear is visible on the left, and the chains of stars around the north pole were conceived as analogous to the walls of the Imperial Palace: that is, in Chinese thought they enclosed the pivotal region of the universe.

months of the year, the Sun's position in each of twelve signs of the Zodiac was formalised. Pegasus's place was more accurately filled by Aquarius, and each sign occupied 30 degrees, corresponding to the thirty days in each month.

This division of the sky into twelve zones, and the ability to plot celestial positions led to a quickening development of astrology in the first millennium BC. The aspects of the heavens could now be formalised: the movements of the planets, the personal deities whose influence was so important, through the Zodiac, and the characteristics of the Zodiac figures could be elaborated into a psychological drama into which the human subject stepped to play his or her part. The generalised predictions of the omens were replaced by the archetypal form of astrology, the horoscope, which derives from the situation of the heavenly bodies at a precise moment, that situation being known and charted long before or long after the moment itself. A complex network of influences involving the entire heavenly sphere, visible and invisible, could now be plotted, clearly a more sophisticated concept than the earlier omen system. It is difficult to believe that such a significant exercise in zoning did not result in the creation of some type of map or model, whether in two dimensions or three, but if they were made in ancient Mesopotamia, they have not survived. From this zoning sprang, later, the most important innovation in man's charting of the sky, the concept of a co-ordinate system to locate heavenly bodies, using the ecliptic and celestial poles as fixed points from which to measure any position in the sky. When and where this crucial step was taken is uncertain; the Babylonians themselves seem not to have used a co-ordinate system as we know it, yet by the third century BC the Babylonian Zodiac was certainly known in Greece, where, plotted within the emerging rules of spherical geometry, it produced the classical form of celestial latitude and longitude which we understand today.

Astronomy Beyond Europe: China – Central America – India

Beyond Europe and the Near East the same imperatives of time-measurement and divination produced quite independent astronomical traditions. In China by 1000 BC the state maintained astronomers to draw up calendars, keep time, interpret omens and monitor weather conditions, and this state-supported science reached a high degree of observational precision. It seems possible from the references of later astronomers that star charts of some sort were

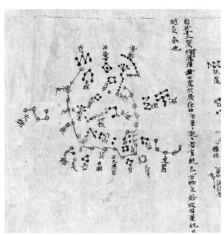

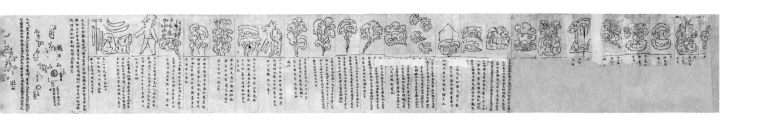

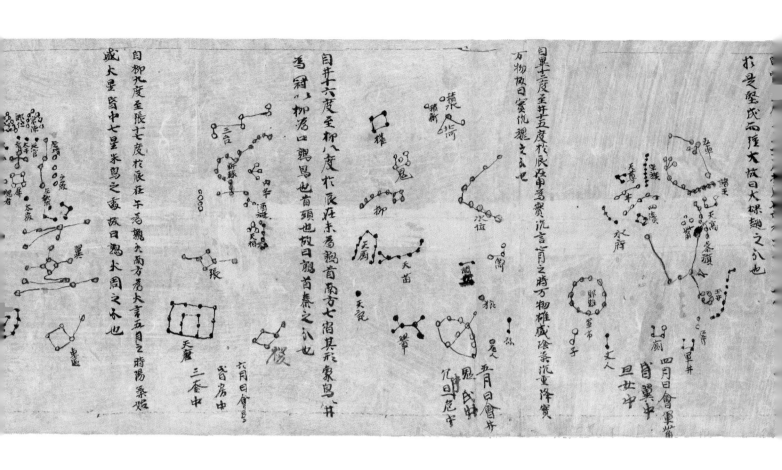

自柳九度至張七度於辰在午為鶉火南方者大言五月之時陽氣猶
歲大星昏中七星朱鳥之象故曰鶉火周之分也

自井十六度至柳八度於辰在未為鶉首南方七宿其形象鳥井
為冠以柳為口鶉昌也首頭也故曰鶉首秦之分也

自畢三度至井十五度於辰在申為實沉言苗之時万物稚藏陰氣沉重降實
万物故曰實沉趣之分也

於曼墾戌而陸犬成曰大樑趣之分也

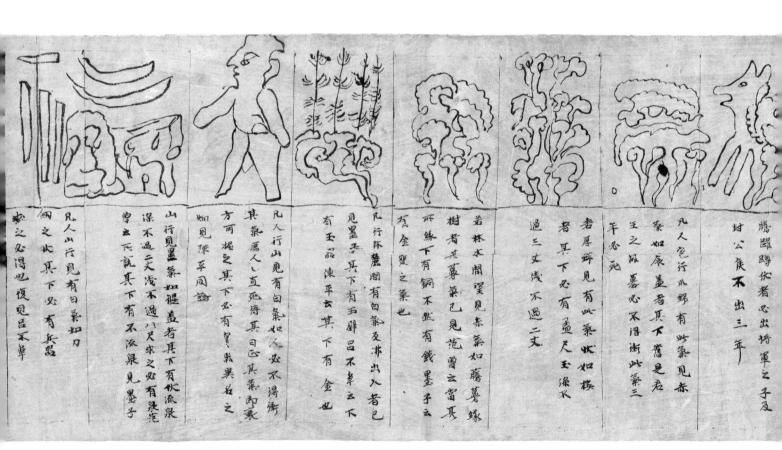

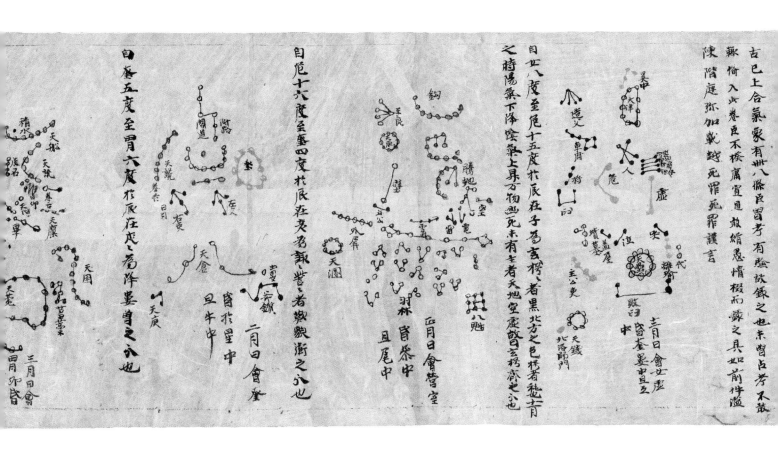

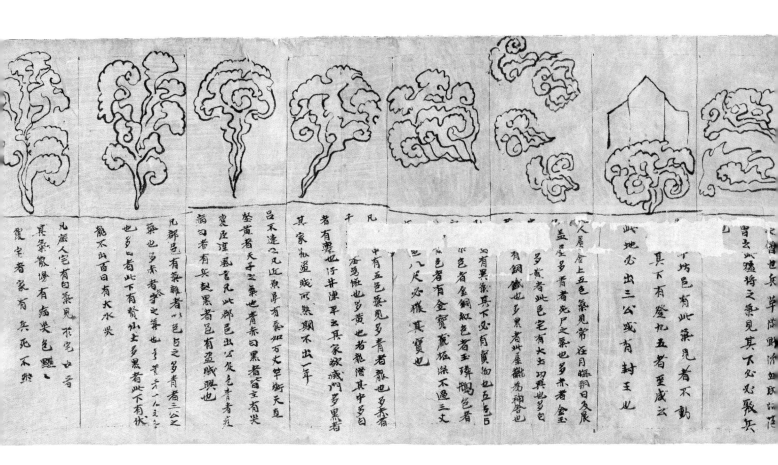

Above
Mexican astronomer. A priest or astronomer is seen watching the seven stars of the Pleiades, *c.* 1300 AD.

in existence in China by the fourth century BC. These have not survived, but a fully worked-out star catalogue listing 1464 stars in 284 constellations was drawn up by Chien Lu-Chih in the fifth century AD. The oldest extant two-dimensional star map from any civilisation comes from the tenth century AD. Found among the cache of manuscripts at Dunhuang on the ancient Silk Route, it clearly belongs within a mature tradition of celestial mapping. On a more philosophical level, there was in China a marked absence of cosmological modelling, and no impulse to people the cosmos with deities and demons. Chinese philosophy apparently did not conceive of a creator-god standing outside his created world, and Chinese science was not drawn to rationalise the laws of nature or causation, whether the causation of physics or of the divine will. Instead the prevailing concern, whether of Confucian pragmatism or Taoist mysticism, was the organic wholeness of the universe. This wholeness could be grasped intuitively through the perception of nature's balanced forces Yin and Yang, manifest in the antithesis of day and night, male and female, Sun and Moon, hot and cold, and so on. Such a 'worship of the universe through the worship of its parts' produced a caste of thought very different from the synthesising rationalism of western science.

By contrast the ancient astronomy of Central America, equally precocious in exact observation, was used to underpin an exuberant cult of deities, demons and cosmic imagery. The Aztec, and before them the Maya peoples, developed the ability to analyse astronomical events mathematically and to predict events with great accuracy. Their calendar-making advanced beyond time measurement into the designation of an elaborate cycle of unique identifiable days, extending over fifty-two years, or 18,980 days, called a Calendar Round, after which it repeated itself. Above and beyond the Calendar Round was the famous long count, which traced the history of the creation (or at least the current cycle of it) back to 3114 BC in the western calendar, although the reasons why this date was chosen are still unknown. Each day was regarded as possessing a certain fateful character derived from the ascendant stars. Personal astrology in the western sense did not develop, but the reading of astral omens became an essential priestly function. Their elaboration in almanacs belongs perhaps to the history of anthropology rather than astronomy, but the central cults of the Sun and Venus (the latter as the male god Quetzacoatl) were underpinned by rigorous observation over many centuries. Highly animistic pictures of the Sun and other gods are common, but there is nothing in the nature of star maps. The Central American cultures had not developed the science of geometry, and spherical geometry in particular, which was the key to western cosmic modelling, was unknown. Instead they seem to have conceived of a flat, layered universe, each layer the domain of one kind of celestial body – clouds, Moon, Sun, stars, comets, planets, and in the final layer resided the creator-god. This system clearly precluded any attempt to grapple with the dynamics of astral motion. It is striking that the Central American cultures, like the Mesopotamian, developed sophisticated mathematical astronomy but left so little graphic record of their science.

In India the ancient Vedic literature, dating from *c.* 1500–1200 BC, is full of references to astral gods, and to the balance of cosmic forces both before and after the world's creation. There is no evidence of mathematical techniques of astronomy in India before the fifth century BC. One distinctive feature of the Indian perception of the sky was a celestial reference system relating not to the Sun but to the Moon – the *naksatras*, the lunar mansions – which was established by 800 BC. In this system it was the position of the Moonrise against the night sky which defined a series of adjacent star groupings. Although simpler than calculating the Sun's position against invisible stars, it was not destined to replace the Zodiac outside India; yet the system had a long and involved history and formed an important feature in astrology, eastern and western. From the fifth century BC onwards, classical Indian astronomy developed under a series of influences from the west: from Mesopotamia, Persia, Greece and, later the Islamic world. Key texts from these foreign sources were translated into Sanskrit, and Indian astronomers embraced the Babylonian Zodiac, Aristotelian cosmology, the Ptolemaic star catalogue and Hellenistic astrology. In the cosmology of the Puranas (popular encyclopedic literature *c.* 400–800 AD) a series of wheels bearing the heavenly bodies revolves above the Earth, their axis in the sacred mountain Meru,

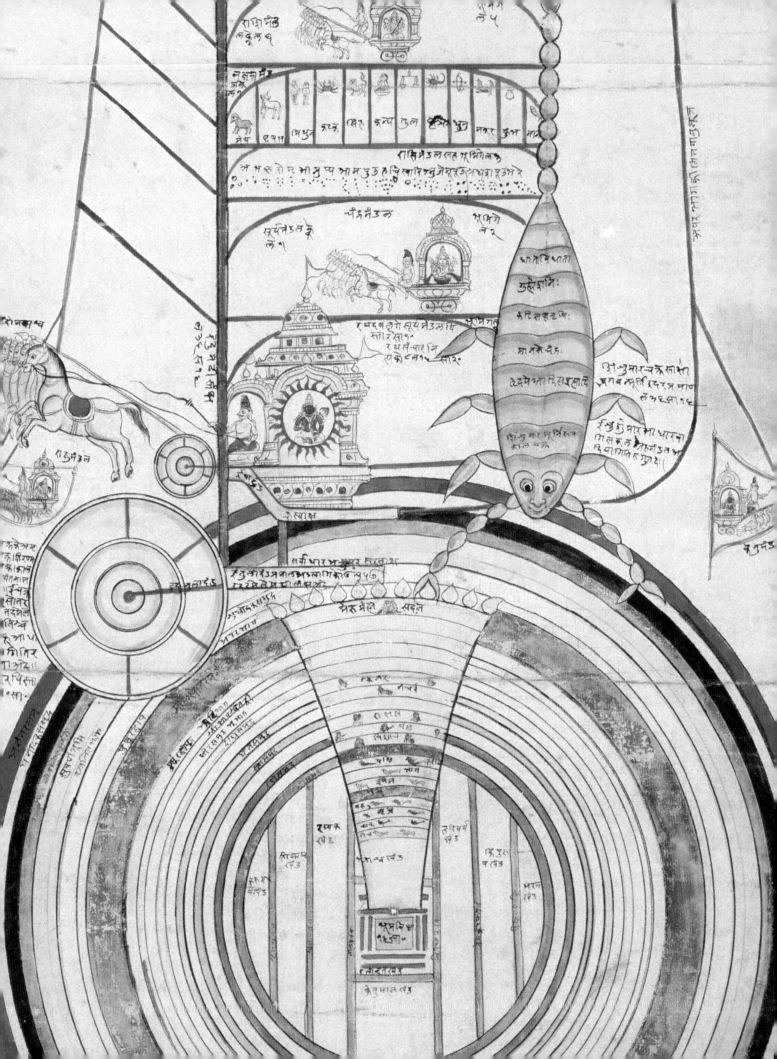

Opposite
Diagram of the paths of the
planets, painted on cotton.
South India, c. 1750.

their motive power the breath of Brahma; this was a vision clearly rooted in Greek cosmology, but not one worked out in terms of spherical geometry. But armed with mathematical techniques from Babylonia and the concept of spherical motion from Greece, Indian scholars proceeded to calculate their schemes of immense cyclical periods of time, the yugas and kalpas, which were related to the positions of the heavenly bodies. In this tradition, actual observation of the heavens played a minor role, and they were content to take astral positions and theories of motion from sources such as Ptolemy. The ingenuity with which they worked out their aeons of times was unique and remarkable, and we could almost think it was designed to prove that time and eternity were precisely the same thing, which sounds like a very modern insight. Indian wisdom was essentially other-worldly, focusing on ways of escape from the cycle of action and suffering and the mechanisms of fate, of raising the mind to eternity, and in this quest astrology too became central. While star maps in the strict sense were unknown in classical India, there was an exuberant tradition of cosmological charting and modelling, especially in the Jain tradition, in which human life was represented in pictures together with the elements of the natural and spiritual realms.

Greek Science from Homer to Ptolemy

Contemporary with the mature phase of Babylonian astronomy – the designation of the Zodiac and the incipient art of astrology – the foundations of a dramatically new and different scientific tradition were being laid in Greece. From the earliest Greek philosopher-scientists whose names have come down to us – Thales, Anaximander, Pythagoras, etc. – there is a decisive shift in approach towards a critical, analytical scientific spirit. The universe is regarded as a rational structure, capable of yielding its secrets to logical thought and inquiry. The will of the gods was no longer adequate to explain the forces of nature and the phenomena which humans observed, and instead secular explanations were sought which satisfied the questioning intellect. Fundamental questions such as the nature of matter were addressed for the first time. The central problem which came to dominate Greek astronomy was that of the structure of the cosmos: what model could explain the motions of the heavenly bodies in relation to the Earth? This had never been addressed as a theoretical problem by the Babylonians or Egyptians, and it was the Hellenic skill with geometry, especially spherical geometry, that explains their progress in this area. It has to be emphasised that no celestial maps or diagrams of any kind have survived from the classical Greek period, but their approach and their achievement are recorded in some detail in a number of important texts, and it seems highly likely that they did make such maps. Moreover, in recent years a small number of ancient celestial globes have come to light, made of metal and showing many of the classical constellations. It is not certain at what date these globes were made, nor exactly where they come from, but they were found in the Near East, and most experts believe that they are certainly Greek, and probably from the late classical period, within the range 200 BC–AD 200. This agrees with what we know of the work of the astronomers Eudoxus and Hipparchus, who contributed to the concept of the celestial sphere, and were said to have constructed celestial globes.

Traditional Greek lore, as preserved in Homer for example, had regarded the sky as a dome of iron or bronze supported by pillars over a static and probably flat Earth. By the fourth century BC, however, Parmenides had satisfied himself that the Earth was a sphere, and that the Moon shone with reflected sunlight, while Empedocles was probably the first to interpret correctly the geometry of solar eclipses. An idea that had a seminal influence was Pythagoras's teaching concerning the perfection of the figure of the sphere, a perfection that was both physical and metaphysical. This was typical of the Greek ability to handle an abstract, almost intuitive concept, and to build logical conclusions from it. For the Pythagoreans, it followed from this belief that the Earth must possess the most perfect form known to nature, the sphere, and that the heavens which surround the Earth must also share this spherical structure. Greek astronomers were also aware that as one travelled north or south, different stars rose into view while others dropped below the opposite horizon. Unless the stars were very close, this could only be explained if the Earth's surface were curved, so that there were good empirical as well as philosophical reasons for regarding both the Earth and the surrounding heavens as spheres.

Later Greek scientists and philosophers all became convinced that spherical motion held the key to the structure of the cosmos. It was on this aspect of the mapping of the heavens – the structural rather than the locational – that Greek science expended so much thought and had such long-lasting influence. It is difficult to overemphasise the importance of the concept of the heavenly sphere: it offered a structure, a plane surface, upon which stellar positions could be measured and mapped according to the laws of geometry. This provided the basis for all serious celestial mapping, and it incidentally provided Greek geographers with a model for terrestrial mapmaking too.

The motions of the stars and planets, Sun and Moon, were each clearly regular, but they were far from uniform with each other. The retrograde motions of the planets in particular offered a long-standing enigma: what model could explain the mysterious pattern of their movements, and what mechanism supported them and the other heavenly bodies in the sky? In the *Republic*, Plato relates the myth of a man killed in battle whose soul travels to the regions of the spirit world. He is granted a vision of the cosmic structure as a nest of concentric hoops encircling a central axis, each hoop bearing a planet. The Earth is a kind of spindle at the centre of the figure, and the hoops are turned by the Fates. In a later work, the *Timaeus*, Plato again describes the universe as a series of circular bands around the Earth. Some scholars have argued that these texts are so graphic that Plato must have had before him a three-dimensional model of the heavens, like an armillary sphere, or even a mechanical model. More than a century after Plato, Archimedes is reliably reported to have built such models. Plato gave expression to the Greek fixation with the sphere: 'Therefore the creator fashioned the world a rounded, spherical shape with the extremes equidistant in all directions, a figure that has the greatest degree of completeness and uniformity ... and he established a single, spherical universe in circular motion.'

It was a contemporary of Plato, Eudoxus of Cnidus, who first articulated a theory of cosmic structure based on circular motion with the Earth at the centre of twenty-six circular bands bearing the Sun, Moon, stars and planets. Why twenty-six ? Because each celestial body has a double or even triple motion – the daily rotation and the yearly cycle – which could be explained only by a spherical orbit which bore upon itself the centre of a second movement. The elaboration of these two elements of the system became central to Greek cosmology. The general approach to cosmic structure advanced by Eudoxus acquired immense historical importance because it was shared by Aristotle and Ptolemy, and with their authority it became the accepted and dominant cosmic model in western thought for almost 1,500 years until the Copernican revolution. Eudoxus's second great achievement as an astronomer was the construction of the first recorded celestial globe, depicting the constellations within a co-ordinate framework of the ecliptic and tropics, the poles and the equator. This clearly drew upon an established corpus of observations, and was influential in fixing the classical constellations, which have remained unaltered ever since. It is certain that Babylonian knowledge had reached Greece by this time, since the Babylonian Zodiac was included.

Like all future celestial globes, the starry sphere is described and envisaged as seen from *the outside*, from some imaginary point in space beyond the cosmos. The practical difficulties of constructing concave or domelike hemispheres to be viewed from inside were obviously too great; but moreover there was no conceptual obstacle for the Greeks in looking at the stars from 'outside space', since the starry sphere was conceived literally as a material but translucent sphere, located at a specific distance from the Earth, on which the stars were carried. It was assumed that each of the other heavenly bodies was also borne on its own crystalline sphere, with the spheres revolving independently inside each other. The depiction of the constellations from the outside had the important effect of reversing the figure-images: the zodiac progresses anti-clockwise around the ecliptic, and the head of Taurus, for example, faces to the right on the globe, but to the observer on Earth it faces left. This became an established convention, and virtually all celestial globes down to the present day have been drawn as if from outside the stars. Eudoxus's globe has not survived, but it was undoubtedly the prototype of a series of Greek celestial globes, and the Farnese Atlas – a marble sculpture of Atlas bearing a celestial globe, which dates from the second century AD – is its direct descendent.

Eudoxus's original text was rewritten in verse around 25 BC by the poet Aratus of Soli, and the resulting work, the *Phaenomena*, became one of the most popular scientific texts in the classical, post-classical and medieval worlds. Since Eudoxus's own work is lost, Aratus's poem is the oldest known

ORBIS CÆ
Ex mar.
in ÆDIB
R

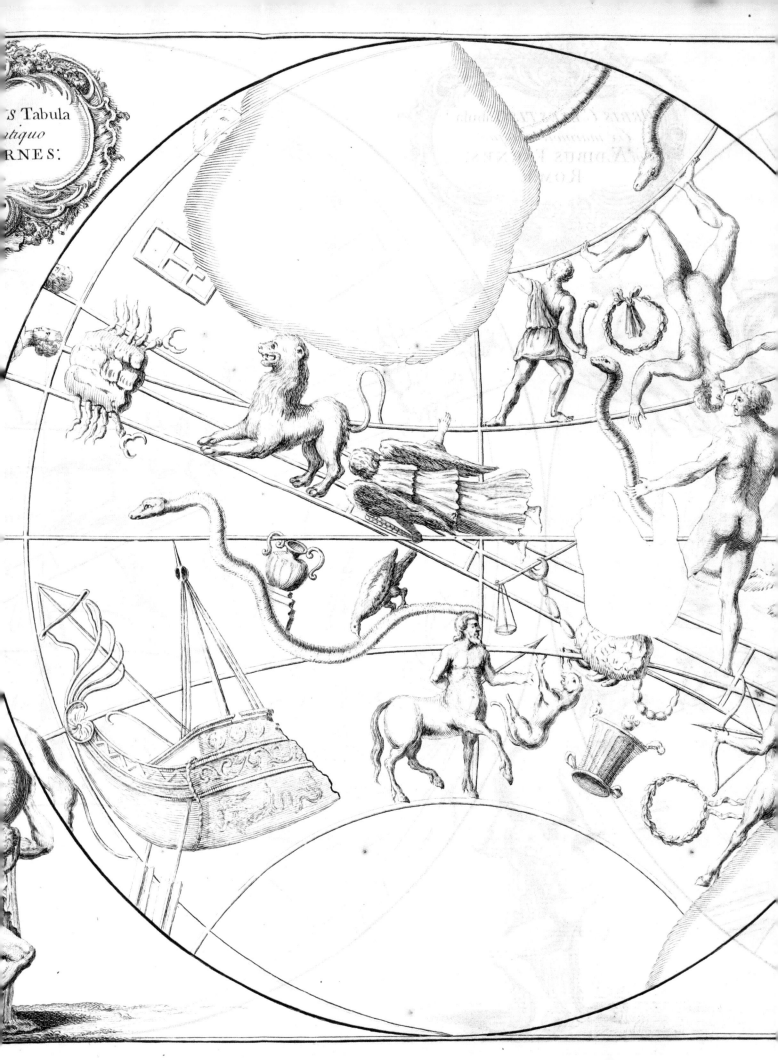

systematic account of the classical constellations. Their appearance and
location are given in some ten to twenty lines of rather rhetorical verse, and
reference is made to the mythical background of the figures. The planets
are also described, and although the cosmic structure is not elaborated in
any detail, a system of geocentric circles may be deduced from what Aratus
says. The positions of the planets in the constellations is used to give a series
of detailed weather predictions, but no personal astrology enters into the
poem. Aratus's poem is a precise verbal description of forty fairly simple
visual images, and it is hard to believe that texts of the poem would not,
from its earliest appearance onwards, have been illustrated with a series of
constellation-pictures.

Aristotle echoed the Greek preoccupation with spheres and circular
motion as marks of perfection in nature. Aristotle divided the cosmos
into the sublunary realm (where all matter was composed of the four
elements earth, air, fire and water), and the celestial realms (where the
ether reigns, a unique fifth element not found on Earth). Linear, irregular
or intermittent motion characterises things on Earth, while the heavens are
qualitatively different; there, only perfectly circular motion is found. This
metaphysical, almost spiritualised, science is further reinforced by Aristotle's
doctrine that all motion is caused, and caused constantly: anything that
moves perpetually, as the Sun or stars do, is perpetually being moved, and
the motions of the heavens he ascribed to a prime mover, this *primum mobile*.
This can be conceived as operating in the outermost sphere of the heavens,
and each inner sphere was somehow 'geared' to its neighbour. By adding
these differential, or counteracting, spheres, which were needed to isolate
each sphere from the motion of the next, Aristotle increased the total number
of spheres to fifty-five. The mysterious prime mover, itself unmoved and
eternal, would later be easily identified with God, so that the Aristotelian
cosmic structure was found acceptable to Christianity for well over 1,000

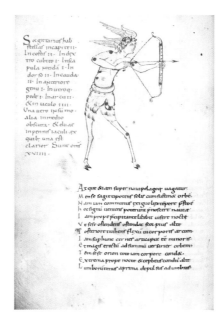
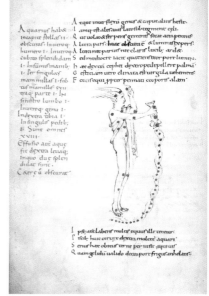

years. Aristotle's authority also reinforced the enduring idea that the Earth is stationary: both our senses and our logic tell us that the Earth is vast and too heavy to move.

The early high-point of this rigorous intellectual science came with the work of Hipparchus (active 150–130 BC), the most important astronomer before Ptolemy. His own works have not survived, but their contents are known from detailed references to him by Strabo and by Ptolemy himself. He pioneered the application of trigonometry to celestial calculations; he developed an early form of astrolabe; he constructed a star globe, and experimented with celestial co-ordinate systems. He understood clearly the relationships involved in eclipses, he used parallax geometry to calculate accurately the distance from the Earth to the Moon, and he appreciated that the Sun was vastly more distant, beyond his measuring ability. His most famous observational discovery was what later came to be called precession: he noticed that star positions are not absolutely fixed, but that the whole starry sphere moves slowly westward around the ecliptic. He was unable to measure this movement with great accuracy, but he put it at around one degree in a century (the true figure is 50 seconds of an arc each year, or one degree in seventy-two years). Hipparchus was not able to account for this motion, but he realised that its effect was to render any star maps or catalogues giving star positions valid for only a certain epoch, perhaps for half a century, depending on the accuracy the observer aims at. Hipparchus's prodigious skill as an observer was summed up in his pioneering star catalogue, listing 850 stars with their co-ordinates, and classified into six magnitudes of brightness. Hipparchus did not elaborate on the prevailing concentric cosmic model of Greek science, but he found no reason to quarrel with it. His approach to astronomy was precise, empirical and mathematical, and it seems certain that he made celestial maps, diagrams and globes for his own use; but, as with classical maps of all kinds, they have perished. For almost three centuries no other astronomer approached Hipparchus's originality and precision.

It was Ptolemy of Alexandria (active 130–160 AD) who both summed up the achievement of classical science, and established the canon of astronomical knowledge for the following 1,300 years. In the introduction to his great book, the *Syntaxis*, better known by its later Arabic title the *Almagest*, Ptolemy advances an interesting, almost religious, plea for the study of astronomy. Moral insight, he argues, can grow from our reflecting on our worldly affairs; sciences such as physics and medicine deal with the changing, corruptible material world; but astronomy alone leads to a deep knowledge of the universe, focusing as it does on the eternal heavenly bodies, the motion of the cosmos, and the divine first mover. Ptolemy revised Hipparchus's star catalogue and expanded it to include 1,022 stars (including the few nebulae visible to the naked eye), whose co-ordinates and degrees of magnitude he gives. These stars and the forty-eight constellations in which they were grouped formed the basic material of all astronomy in the western world until the early seventeenth century. More dominant even than Augustine in theology or Virgil in literature, Ptolemy's authority as the guide to all things

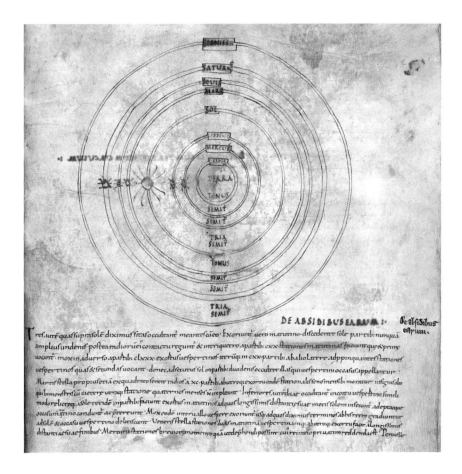

astronomical was paramount throughout the medieval and Renaissance world, and the *Almagest* was probably in continuous practical use longer than any other book in history, with the single exception of Euclid's geometry.

The concept of plotting positions on the celestial sphere by means of co-ordinates, as it was developed by Eudoxus, Hipparchus and Ptolemy, was transferred to the earthly sphere, and formed the basis of the late classical theory of mapmaking. No less important was Ptolemy's restatement of the circular cosmic model. In great geometric detail, Ptolemy calculated the motions of the planets, Sun, Moon and starry sphere, each with interlocking cycles to represent daily and yearly revolution. He went further than his predecessors in stating that the universe could contain no empty space, and that the sphere of each heavenly body could not overlap another. He was able to calculate the distance of the Moon from the Earth, and from this he built, step by step, a tentative picture of the total size of the cosmos. Ptolemy expressed his estimates in multiples of the Earth's radius (ER), which he knew to be around 5,000 miles. If we convert these ER numbers to miles, we learn that he considered the Moon to be about 300,000 miles distant; the Sun was 5 million miles; Saturn, the most distant planet in the solar system, was 80 million miles, and the sphere of the fixed stars was some 100 million miles away. Thus the diameter across the entire spherical cosmos, of which the Earth was the centre, came out at 200 million miles. This figure seemed sufficiently awe-inspiring to be commensurate with the wonder of the heavens, but without raising disturbing problems connected with infinity: the universe was vast certainly, but not too vast. Thus he portrayed the universe as a finite, closed system, a vision which satisfied both the intellect and the faith of Europe for more than a millennium. His estimates of the absolute dimensions of the cosmos would not be seriously challenged until late in the sixteenth century.

The *Almagest* gives specific instructions on making a celestial globe, but no mention is made of two-dimensional star charts. In another work however, the *Planisphaerium*, he describes the polar stereographic projection which underlies the astrolabe (see page 71). It is theoretically possible that the instrument was known to the Greeks of this period, although no material evidence has survived. Ptolemy warns that the problem of precession will in time render a star globe inaccurate, so he advises that the celestial equator and central meridian be not marked on the globe itself but on detached bands mounted around it, so that the globe itself can be rotated within the co-ordinate framework. This instruction was almost universally ignored by later Islamic and European globe-makers. The stars listed in Ptolemy's catalogue are given a number within each constellation: Aldebaran (not so named) is number 14 of Taurus, 'the bright star of the Hyades, the reddish one of the southern eye'. Betelgeuse, brightest star in Orion, is 'the bright red star on the right shoulder'. To our eyes these textual descriptions could easily be replaced by graphic images, and it is hard to imagine that Ptolemy and his scribes were not motivated to draw constellation diagrams or maps; but if they were made in antiquity they have not survived. Yet in the light of the Antikythera mechanism, dating from around 100 BC and discovered early in the twentieth century, it is certain that Greek scholar-scientists had the means to design graphic models of the heavens, and it appears inconceivable that charts of the heavens were not also produced. With its plates and cogwheels, this mechanism has been variously interpreted as a calendar calculator, or even as a working model of the heavens, rather like an eighteenth-century orrery. In either case it has clarified our view of the way Greek thinkers saw the structure of the heavens, and how they were able to create accurate models of it. It is possible that the Antikythera mechanism is a unique chance survival of the kind of model that Archimedes was reported to have made.

Astronomy and the Philosophical Quest

In the elaboration of cosmic models and the construction of the first star maps and globes, we see spherical geometry being used in the service of Greek analytical thought. The plotting of stellar positions on three-dimensional star spheres and the conceptual modelling of planetary systems were both evidence of the secular theoretical probing of natural laws so characteristic of Hellenic science. Yet alongside this rigorous intellectualism there flowed an altogether different, a darker, more equivocal stream of astronomical thought. For all the progress in precise observation and geometry, the metaphysical question still remained: What *were* the stars and planets , and how did they

Above
The Antikythera mechanism. Found badly corroded in an ancient wrecked ship, this mechanism was in some sense a working model of the heavenly spheres, possibly dating from around 100 BC. It may have been used for calendar calculations, and it has revolutionised our view of ancient Greek technology. It seems inconceivable that it could have been designed and made without using carefully drawn maps of the celestial sphere as its basis.

Above
Scorpio and Ophiuchus from a
fifteenth-century manuscript of
Aratus. We do not know whether
classical manuscripts of Aratus
and other astronomical works
were illustrated or not, and we
remain dependent on medieval
reconstructions. This picture is
typical of medieval constellation
imagery, in which astrological
interest focused attention on the
qualities of the personified stars
and planets.

relate to human life? If the cosmos was a closed, finite system, how did
it cohere? What was the prime mover, and what was man's place in this
flawless but mysterious mechanism? Behind the empirical pursuit of scientific
knowledge, some minds would turn to the search for cause and meaning, the
coherence beneath the observed phenomena, which must surely come from
some source beyond the physical realm. Plato had been in no doubt that 'the
fixed stars are living beings, divine and eternal' and that the heavens are 'a
moving image of eternity'. It was the elaboration of such beliefs that guided
the astrological enterprise, which provided a further powerful motivation for
the study of astronomy.

The central doctrine of ancient astrology was that the heavenly bodies,
and above all the planets, shaped earthly events by the exercise of divine
powers, which in turn must mean that the planets were themselves divinities,
living entities. This was the doctrine on which the Babylonian omen-culture
was founded, and it can be seen in the Babylonian names for the planets,
which were those of the gods: Ishtar-Venus, Marduk-Jupiter, Nergal-Mars
and so on. Archaic Greece, on the other hand, had no system of astral
religion; for example, in the Homeric poems we find no personal planet
names. But the transmission to Greece of the idea of planetary gods is evident
by the time of Plato, when Jupiter was named 'the star of Zeus', Mars 'the
star of Ares', Venus 'the star of Aphrodite' and so on. After this, the Romans
in turn gave the planetary deities the names by which we know them today:
Jupiter, Mars and Venus.

It was believed that the gods' power waxed and waned according to their
positions in the heavens, for example in relation to the Zodiac signs, and
these positions and their effects could be charted and analysed. This was a
highly sophisticated form of divination, and clearly springs from the belief
that the gods habitually gave humans signals of their intentions: they wrote
their will on the face of nature, and especially in the heavens, where they
dwelled. The ancient experts learned, or claimed, to interpret these signals
by building up a precise science of astronomical positions. The universe was
conceived as a complex of spheres and forces centred upon the Earth, and it
became the aim of astrology to chart the balance of those forces at any given
moment. Astral religion, the belief that personal divinities inhabited the
planets, spread from Babylonia to Persia, India and Egypt and throughout
the Graeco-Roman world. It carried astrology with it, to the extent that
this became an important component in the intellectual life of the entire
Christian and Muslim worlds, a system that was half-way between a religion
and a science.

In time, the focus of interest settled on the vitally important moment
of a person's birth, with the aim of revealing an underlying pattern which
would determine the nature and destiny of that individual's life. All the
elements in this system – Sun, Moon, planets and constellations – could be
disposed in an infinite variety of ways, and moreover they were believed to
be linked to the earthly elements of air, fire, water and earth, and to animal
and natural characteristics, to create each individual soul. This interweaving
of the human and cosmic material became articulated as the macrocosm–

microcosm link, which became a ruling principle in the astrological
enterprise. The key point to notice is that for astrology to develop into a
credible science, for the horoscope to be cast, it was dependent on precise
observation of the celestial sphere, on precise zoning and charting of the
heavens. It progressed on the basis of strict astronomical knowledge, which
was used to plot astral positions not only in the present but back in time and
forward in the future.

A major change in the rationale of astrology came with Ptolemy, the great
astronomer who also wrote a seminal textbook on astrology, the *Tetrabiblos*,
which summarises a vast body of Hellenistic lore. Ptolemy is guarded on the
question of the personality of the celestial bodies, and he expresses the view
that the influence of the stars and planets on human life stems entirely from
the balance of physical forces in the universe, rather than from their power
as divinities. In other words, the planets exercise a natural physical influence,
and are no longer seen as gods or goddesses. He elaborates this view by
describing the meteorological effects that flow from the various conjunctions
of planets and constellations, following the model of Eudoxus and Aratus. In
the same way, human life and human events are subject to physical influences
from the heavens. No one would deny that the Sun and Moon affect human
life, and Ptolemy argued that the other heavenly bodies must do so too, only
at a more subtle level. The disposition of the stars and planets is seen as an
objective force shaping the human personality, obedient to laws which may
be discerned and charted, but not ultimately explained. These concepts were
later to become almost universally accepted in both Islamic and western
astrology. This was a theory that was more natural and scientific than the
astral religion which saw the planets as living gods. The fact that a major
figure such as Ptolemy wrote an extended treatise on astrology shows that
there was no rigid demarcation between astronomy and astrology.

The Babylonian system of astrology, including the planetary gods and
the all-important Zodiac zoning, had evidently spread through much of
the ancient world by the fourth century BC. When in turn Alexander the
Great conquered this entire realm, the way was open for an unprecedented
exchange of knowledge and beliefs in the fields of science and philosophy,
and a complex syncretism of religion and magic. Not surprisingly, the precise
sources, developments and interrelationships of these beliefs have not been
fully worked out. But throughout the Hellenistic world, astrology became one
of the acknowledged pathways of religious and intellectual search. It rested
securely on astronomical skills and the astrologer was often, but not always,
a scientist and philosopher too. It was in this period that the terms Chaldean
and Magus testify to the eastern unity of astronomy and occult knowledge,
and it is from this period that the earliest extant Zodiac survives, the Zodiac
from the Dendera Temple. Although this carving is from the Egypt of the
first century BC, it undoubtedly embodies Babylonian concepts and figures,
married with the traditional figures of the decans.

For some reason divination by the stars and planets, with the
interpretation of omens, was regarded by the Greeks as native to Egypt,
which suggests that Babylonian traditions reached Greece via Egypt.

Herodotus described what was evidently a fairly mature astrological practice when he visited Egypt in 460 BC:

> I pass to other inventions of the Egyptians. They assign each month and each day to some god: they can tell what fortune and what end and what disposition a man shall have according to the day of his birth. This has given material to Greeks who deal in poetry. They have made themselves more omens than all other nations together; when an ominous thing happens they take note of the outcome and write it down; and if something of a like kind happens again they think it will have a like result.

This is a fascinating account of ancient astrology, but what Herodotus is describing here is Babylonian astrology and omen-lore, not Egyptian. At this date the Egyptians did not possess sufficient skill in plotting stellar and planetary positions to enable them to cast horoscopes. Herodotus's tone implies a healthy scepticism, a reminder that at this date these mystical beliefs were novel and not yet as readily comprehensible to many Greeks as they would later become.

While empirical astronomy devoted itself to gathering objective data, and building theories from it, astrology sprang from the subjective conviction of the organic unity of the heavenly and earthly worlds. One of the central and recurring beliefs in the ancient world that underpinned this search for unity was the doctrine that the original home of the human soul was in the stars. Pythagoras apparently taught that the soul fell to the Earth and became housed in the body, and from this it followed that the object of humans' spiritual striving is to liberate themselves from this world and return to their celestial home. Plato echoed this doctrine, and it provided rich source-material for later astrologers, who elaborated on the soul's fall through the many heavenly spheres, acquiring as it fell various characteristics – aggression from Mars, greed from Mercury, lust from Venus, and so on. After death, these qualities would be discarded as the soul retraced its path. To the Hellenistic and then the Roman mystery religions this vision of the descent of the soul was fundamental.

The mystery religions flourished for five or six centuries before Christianity became dominant, and they were undoubtedly a symptom of the quest for a mystical dimension to life denied by the formal state pantheon. The mystery rites themselves re-enacted the soul's progress through death and rebirth, the return to heaven. The Greek cosmological model – the ascending spheres – became central to these religions, and it became grafted onto various oriental cults such as those of Isis and Mithras. The Persian Sun-god Mithras is usually portrayed sacrificing a bull, whose significance is uncertain, but whose blood and seed were the origin of life on Earth. At the moment of sacrifice, the cloak of Mithras was metamorphosed into the sky, stars and planets, while the bull itself became the Moon, suggesting echoes of ancient Near Eastern creation myths. In the Roman period the institutions and rites of Mithraism were undoubtedly interpreted in Neoplatonic terms, focusing on the ascent of the soul after death to the realm of the stars, many of these ideas having their source in Plato's *Timaeus*. Perhaps the height of oriental cult-influence in the west occurred as late as the third century AD when the Syrian Sun-god Sol came near to dominating the official Roman pantheon. Sanctuaries to Sol and the planetary gods proliferated throughout the Empire, and the Emperor Constantine wavered for a time between adopting Sol or Christ as the new deity of the Empire. The festival of Sol on 24–25 December, the winter solstice, became associated with the mystery cults of dying and reborn gods, since after it the days lengthened as the Sun regained its power.

Astronomy and cosmology lay close to the heart of Hellenistic and Roman religion, and the belief that human events are determined, or at least influenced, by the stars was universal. The pragmatic Roman statesman Cicero thought it madness to deny that the stars and planets were personal deities, and in his *Dream of Scipio* he echoes Plato's vision of the soul being granted a glimpse of the cosmic spheres, a vision which re-appears in Dante in essentially the same Ptolemaic form. The problems of fate, determinism and free will thrown up by astrology were not lost on contemporaries. The prevalent secular philosophy of Stoicism taught a calm acceptance of the divine will, a life in conformity with the cosmic order, and the divine will in this context meant the pattern of human fate that was legible in the stars. Part of the seductive power of the mystery religions was their offer of escape from this helpless determinism. A person who progressed through the mysteries could step out of the pre-determined path and shape his or her own destiny. The presiding deity of the mystery cult could unravel the decrees of fate and accelerate the adept's spiritual progress. In this relationship to the god there was thus a foreshadowing of Christian ideas of grace and salvation. The physical, cosmic framework of salvation was always the ascent of the soul to the heavens. The unswerving identification of the skies as the source, the guiding power, and resting place of the human soul was common to all cultures: it gave to the science of astronomy a distinct resonance, and to its practitioners the lure of supreme intellectual adventure.

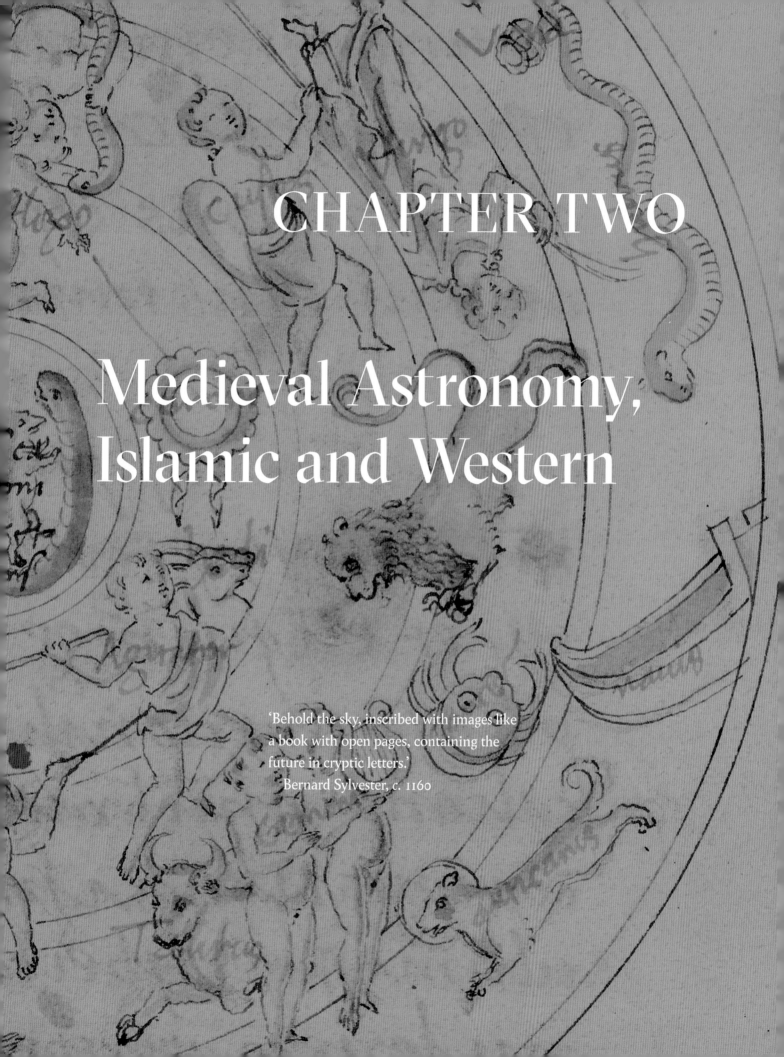

CHAPTER TWO

Medieval Astronomy, Islamic and Western

'Behold the sky, inscribed with images like
a book with open pages, containing the
future in cryptic letters.'
Bernard Sylvester, *c.* 1160

By the third century AD, astronomical thought and practice had reached a historic plateau. The Greek sector of the Roman Empire produced no new figures of the stature of Hipparchus and Ptolemy, capable of advancing the observational or theoretical aspects of the science. In the west, the Romans' distaste for speculative science has often been commented on: new cosmological thought was beyond their reach, and, even on the more pragmatic level, they had enormous difficulty regulating their calendar. With the disintegration of Roman power, the preservation of Greek science in textual form, if not as a living tradition, fell to the scholars and scribes of Byzantium. The nascent Christian church was suspicious or hostile towards astronomy because of its identification with astrology: astrology was one of the manifestations of paganism which the church set out to discredit. The mystery religions and the more philosophical forms of late paganism such as Neoplatonism both had strong natural affinities with astrology. Neoplatonism especially had at its heart the conviction that there were many levels of being, and that this world is a mere image of a higher archetype. There were many possible pathways to the higher realms of being, and the Platonic doctrine that the stars were divinities or intelligences among whom the human soul had its true home, meant that astronomy/astrology was seen as a science through which the chain of being, the harmony of human life with the cosmic order, could be studied and understood.

The Post-Classical World

In some ways the Neoplatonic vision was attractive to the more philosophically minded Christians, but its openness to astrology was anathema to the orthodox early church fathers, whose faith was in a real personal God who acted in human history. Astrology appeared to them to dissolve God's power into impersonal cosmic forces, while the belief that the course of human life might be foretold from the stars apparently denied free will to humans. The near-universal belief in astrology forced Augustine and the other theologians of the early church such as Origen to grapple with this problem of fatalism, and they arrived at a formula much repeated ever since, that the stars incline but do not compel: the pattern of the heavens indicates possibilities not certainties. The striking thing about such a compromise is that it does not seek to deny the basic validity of astrology: celestial influence was accepted as valid in the ancient and medieval world, by both pagans and Christians, but it was seen by Christians as uncertain and dangerous, perhaps even diabolic, and not a fit subject for Christian speculation. Nevertheless it would remain possible for some Christians to think of astrology as one of the

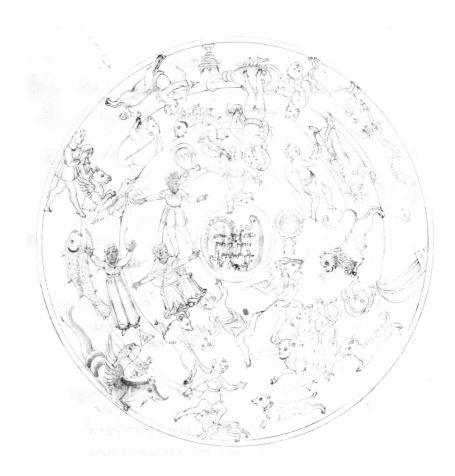

ways in which God's purpose may be discerned. A few of the church fathers, including Clement and Origen in Alexandria, were open to its daring and high-flown speculations, but the Latin Church strongly disapproved of them, and condemned them as heretical.

The years AD 300–800 represent something of a dark age of astronomy, and of science generally. Most of the classical legacy was contained in texts written in Greek, and the complete absence of astronomical tables in Latin meant that the serious practice of astronomy became nearly impossible. It is probable that Aratus's book on the constellations, *Phaenomena*, was known in Greek among eastern scholars and in Cicero's Latin translation in the west, and that copies of it may have been illustrated with rather crude pictures of the star groups. Aratus was certainly the best-known astronomical text from late Roman to early medieval times, but it was distinctly non-specialist and non-technical. A second and similar Latin text was the first-century *Poetica Astronomica* attributed to Gaius Julius Hyginus, which was frequently copied and illustrated during the Middle Ages, and survived well into the era of printing. Hyginus was less concerned with astronomy than Aratus: he does not describe the appearance of the constellations, but recounts the legendary stories that lay behind them. Together, the illustrated Aratus–Hyginus texts formed the basis of all popular astronomy in the late classical and medieval period; they embodied a tradition which was rooted in poetry and mythology, while the Ptolemaic tradition had been far more scientific.

A few highly significant pieces of archaeological evidence survive from this obscure period. The ruined palace of Qusayr Amrah in east Jordan was built about the year AD 715 for the ruling caliph, and although an Arab palace, its site had recently been conquered from the Byzantine Empire where classical influences still reigned. One of the rooms in the palace has a domed roof painted to represent the vault of heaven. Although it is in a poor state of preservation, a number of classical constellations are recognisable, centred around the north celestial pole, while lines evidently representing a simple type of locational system are visible. The constellation figures are drawn anti-clockwise, as if viewed from outside the starry sphere. This mirror image of what the observer on Earth sees is exactly the pattern found on the globe depicted in the Farnese Atlas sculpture, and sources in classical literature are strongly suggested by the Qusayr Amrah dome.

Less than 200 miles from Qusayr Amrah, excavations at the city of Sepphoris, once the capital of Galilee, have revealed another remarkable link, this time between Roman and Jewish culture. After the destruction of the Jerusalem Temple in AD 70, Sepphoris became one of the most important centres of Judaism until the seventh century. The remains of a synagogue have been uncovered to reveal large intricate mosaics in which classical and Jewish subjects are combined. A large Zodiac naming the months and astrological signs in Hebrew has been found alongside depictions of the Ark of the Law and the Ten Commandments. The Sepphoris region was also conquered by Islamic forces in the seventh century, and these two sites thus form striking visual links between classical and Islamic astronomy. For the

Below, left & right
Aratus's *Phenomena* in Cicero's Latin translation, from a tenth-century manuscript. Aratus and Hyginus were the popular sources of astronomical lore throughout the Middle Ages. Their close association with each other is strikingly embodied here in the use of Hyginus's stories to fill in the body of the constellation figures, while Cicero's texts appear beneath the pictures.

renewed pursuit of astronomical science and a renaissance in astrology took
place not among the direct heirs of Greek and Roman thought, but within
the dynamic new faith of Islam.

Islamic Astronomy

After the conquests with which Islam established itself, the eight and ninth
centuries AD witnessed a precocious flowering of cultural activity. Since
their faith's source lay outside the great centres of classical culture, both the
princes and theologians of Islam set out deliberately to enrich their religion
and their cultural life with the best elements drawn from classical science
and scholarship. Centres such as the Abbasid court in Baghdad sought out
texts and teachers of mathematics, philosophy, art and technology, culled
from India, Persia, Egypt and Greece. The fundamentals of what we now
call Islamic science were swiftly synthesised from many diverse sources,
although it later acquired characteristics of its own, and the *lingua franca* of
Arabic gave it a universality that reached from Spain to Persia. The works of
Euclid, Aristotle and Ptolemy were all translated into Arabic by AD 850 and
underpinned Islamic science and philosophy.

Moreover, there were specific motives which stimulated astronomy in particular, the most obvious being the need for an Islamic calendar, to be dated from the *Hijra* of Muhammad (AD 622). Then there was the need to calculate the times of prayer required by the faith, and to locate the sacred direction – the *qibla* – of the shrine in Mecca. These two canons of time and direction engendered an elaborate study in their own right, and they were obviously dependent on precise observations, and on theoretical techniques of astronomy and geography. Medicine and human health, too, were regarded as closely connected with the aspects of the heavens, and every great figure had his physician.

One of the first manifestations of the new science was the appearance of a new type of table – the *zij* – in which celestial times and positions of Sun, Moon, planets and stars were tabulated, with many of the mathematical aids for their calculation. In Ptolemaic astronomy, a complex geometry involving the paths and periods of the planetary orbits had been described, and this had to be applied in order to calculate any given celestial position. In the *zijes*, these positions were set out in tables for a given period of time. The availability of such tables was obviously a precondition for the widespread practice of astronomy. Many of the *zijes* were necessarily valid for one particular latitude only, and several hundred distinct families of *zijes* were produced in centres from Toledo to Samarkand. The phenomenon of precession gave them a limited life, and they had to be regularly recalculated. Some were overtly intended as tools for casting horoscopes and other astrological uses. Later they were to be one of the vehicles by which Islamic skills reached the west as they were translated into Latin in the twelfth century, when they were influential in bringing Arabic numerals to Europe. Their method of construction was securely based on Ptolemy's *Almagest*. Perhaps little that was both new and fundamentally important was added during these centuries, yet on a practical level the Islamic astronomers created the first recoverable phase in the history of genuine star mapping. This took the form not of two-dimensional paper maps but of a remarkable series of instruments – astrolabes and celestial globes – which demonstrated a mastery of astronomical theory and practice far in advance of the science of Christian Europe.

The astrolabe was a hand-held instrument, and most were 6–12 inches in diameter. It consisted essentially of two flat metal plates designed to be superimposed, the one above the other. The top plate was fretted into an open framework bearing a number of precisely placed pointers. These pointers are really a map, giving the positions of several important stars, as if viewed from the north celestial pole. The starry sphere is envisaged as flat and outspread to a latitude below the celestial equator. The small instruments might locate only eight or ten stars, while the larger ones could show as many as fifty. This upper plate was called the 'rete', meaning simply a net. Tangent to the outer edge of the rete was a smaller eccentric circle which represented the ecliptic, on which the Zodiac constellations were marked. On the lower plate was incised a system of celestial co-ordinates, with lines of latitude and azimuth, and most importantly the horizon for the required latitude where

the instrument was to be used. When the rete was placed over this plate, the positions of a couple of prominent stars could be fixed, and the rest of the heavens could then be located. Of course the astrolabe was valid only for a given latitude, but it was not restricted to a particular time of the year, since the rotation of the rete over the horizon plate brought into view the stars of any given season. By separating the objective map of the entire northern heavens from the co-ordinate network which gave their actual positions in space and time, the astrolabe became an instrument both conceptually satisfying and immensely practical. It was equipped with sighting devices to determine the elevation of celestial bodies, and often had interchangeable base plates for different latitudes. When it was used in conjunction with Ptolemy's star catalogue, many hundreds of stars could be identified, and calculations of position, motion and time could be made. The rete was constructed – that is, the stars were plotted – by following an algorithm set out in a number of instructional texts on the astrolabe. The star co-ordinates are plotted in what is technically a polar stereographic projection, having the south celestial pole at the centre, with the celestial sphere down to the tropic of Capricorn outspread on a plane. The classical source of this procedure was Ptolemy's *Planisphaerium* text, and it is in effect a blueprint for a celestial map. Whether such two-dimensional maps were also drawn in manuscript is an open question: if they were, none have survived.

Although it was theoretically possible, it is not certain whether astrolabes were actually made in classical times, but in the light of the Antikythera mechanism (see page 58) it seems possible that they could have been. Also described by Ptolemy, and undoubtedly known and used by him, was the celestial globe, which was revived to great effect by Islamic astronomers. The globe as a conceptual model is much simpler to our eyes than the astrolabe, but it was more difficult to manufacture and far less portable. It was less practical, too, in the sense that what it showed was an abstraction which no human eye could ever see, while the astrolabe was geared to the observer's actual field of vision. The globe was not an observing instrument, and hence was not restricted in latitude nor, if correctly made, in time. The fact of precession gave the astrolabe an accurate life of 50–100 years at most.

One further instrument was less common but relates to the globe and the astrolabe: this is the armillary sphere, in which a small globe representing the Earth is mounted at the centre of a series of metal bands which mark the celestial equator, ecliptic, tropics and polar circles. If sighting points were marked on the bands it could be used to measure celestial co-ordinates. If the starry sphere were added around the bands, clearly the armillary would be transformed into a globe. Conversely if one can imagine the armillary to be perfectly flattened around the north pole, without distortion, the result is a northerly stereographical projection of the heavens, as used in the astrolabe. This perception lies behind the apocryphal story related by a medieval scholar that Ptolemy invented the astrolabe by accident: out riding one day and carrying a celestial globe, Ptolemy dropped it and his horse trod on it, the flattened instrument inspiring Ptolemy with the concept of the astrolabe! In the west the armillary was to become the recognised visual symbol of astronomy, and indeed of science in general, appearing in innumerable manuscripts and printed works, often in the hands of Urania, the muse of astronomy.

The development of these sophisticated instruments was such that in the manuscript texts of all the important Islamic astronomers, no conventional two-dimensional maps appear: they were simply not needed, since anyone studying such a text would also possess these instruments. These same texts reveal the many levels of creative astronomical thought which flourished alongside the practical skills to which the instruments testify. Of the dozen or more great Islamic astronomers of the period AD 800–1200, none was more famous in the Christian west than Abu Mashar (AD 787–886), known to Christendom as Albumazar. He was primarily an astrologer, and used sophisticated arguments drawn from Plato and Aristotle to deflect orthodox religious attacks on his art. In his *Book of Revolutions of the World-Years* and *Book of Conjunctions* he sought to demonstrate that human history, the rise and fall of principalities and powers, coincided with major conjunctions of the planets. Most sensational of all was his doctrine that the world had been created when the seven planets (that is, the five classical planets, the Sun and the Moon) were in conjunction at the first degree of Aries, and that the

Below
Pages from a copy of Al-Biruni's astronomical canon, al-Qanun al-Mas'udi (Baghdad, 1174).

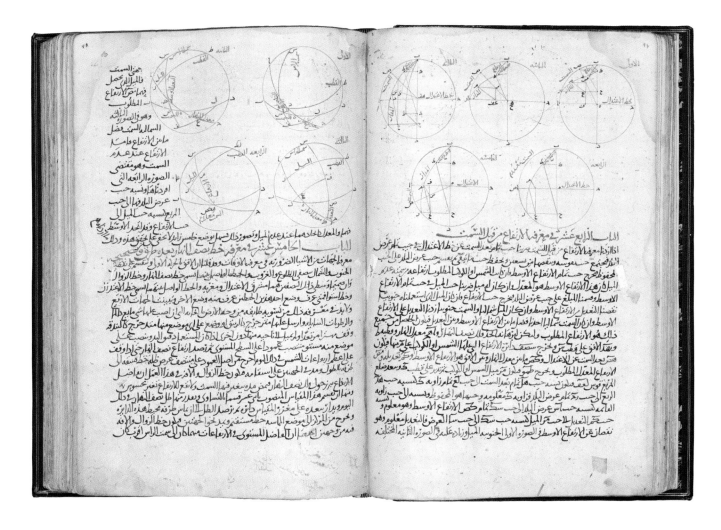

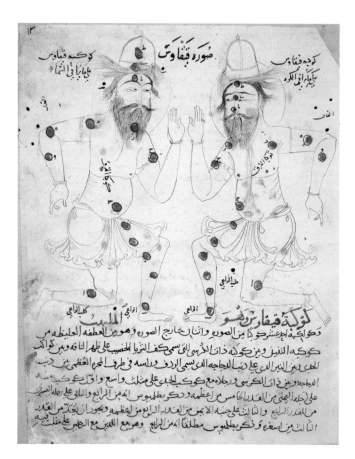
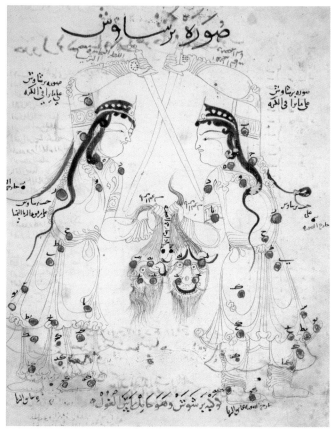

Above

Above

Al-Sufi, *Book of the Fixed Stars*, from a thirteenth-century manuscript. The work of Al-Sufi (tenth century AD) was the most influential of all Islamic astronomical texts. Based closely on Ptolemy's *Almagest*, it catalogued and located more than 1,000 stars, and the manuscripts were habitually illustrated with pictures of each constellation, many shown in double aspect, as seen from the Earth and seen on a globe.

world would end when they all reached conjunction again in the last degree of Pisces. Fortunately unverifiable, this prophecy was widely brooded upon by astrologers of the Middle Ages, Muslim and Christian. Critical of the wilder speculations of Abu Mashar was the equally encyclopedic Al-Biruni (AD 973–1048) who wrote extensively on astrology and cosmology, but who also explored in detail a variety of methods for projecting the celestial sphere into two-dimensional maps. This is puzzling since no such finished maps exist in any extant Islamic manuscripts. Although most of his writing was astrological in content, Al-Biruni's works, including one comparing calendars from different cultures, show an impeccable scientific basis. Yet in contrast with the more colourful Abu Mashar, his works were virtually unknown in Christendom.

The one Islamic astronomer whose fame rivalled that of Abu Mashar was Al-Sufi (AD 903–986), whose *Book of the Fixed Stars* was illustrated with eloquent miniatures of the constellation figures, and was widely imitated in Islamic astronomy. It in turn influenced the iconography of early western star charts. Arabic astronomy from the pre-Islamic age had its own traditional images of the constellations, quite different from the classical figures: a giant human occupied both Orion and Gemini, and a water-jar or bucket covered Aquarius and Pegasus. By contrast, Al-Sufi's work was based almost exclusively on Ptolemy's star catalogue, with the celestial longitudes augmented by 12 degrees to allow for precession to the year AD 964. The drawings show the principal stars graded into six different magnitudes symbolised by size. Not all astronomers were content simply to revise Ptolemy's listings to allow for precession. Painstaking original observation provided the basis for genuinely new catalogues, such as that of Ulugh Beg, made in Samarkand in AD 1437. Ulugh Beg was the grandson of the great Timur (Tamberlaine), and he built an observatory in Samarkand with monumental sighting instruments. These kinds of architectural instruments continued to be built in Islamic countries as late as the eighteenth century, while astrolabes and celestial globes remained in use virtually unchanged for 1,000 years down to the nineteenth century. Islamic science had a deeply conservative aspect, and the new southern constellations observed after the sixteenth century were never added to their maps of the celestial sphere, presumably because they were irrelevant in practical terms.

In the field of cosmology, it cannot be said that there was a characteristic, orthodox Islamic view of the structure of the universe. There was within early Islam a strongly philosophical impulse which sought to provide a rational framework for faith, and this school embraced many Greek concepts. The Ptolemaic planetary system was widely known and became perhaps the accepted cosmology among Muslim scholars. Abu Mashar, for example, expounded the theory of astral influence as working through the physics of the Ptolemaic rings. But, as in medieval Christendom, there were several more mystical, even Gnostic, schools of theology, who claimed to discern the hidden or inner truths of the creation and the cosmic structure. The texts expounding these doctrines were often accompanied by diagrams showing a plurality of human and cosmic elements. These diagrams share perhaps two main characteristics. First, they unfailingly employ regular geometric shapes, especially circles, clearly echoing the Greek sense that the circle and the sphere are nature's most harmonious shapes. Sometimes the whole structure is a series of concentric circles; sometimes it is a group of adjacent or interlocking spheres. Secondly their motive is clearly to join intellectual and material categories in idealised relationships, to express an essential harmony that is invisible to the eye. In all these pictures, planets and Zodiac constellations are part of the hierarchy of being, and major figures such as Al-Biruni contributed to this tradition.

During the great period of Islamic astronomy, AD 900–1200, both observational techniques and theoretical mastery were in advance of those in Christian Europe. It was the translation of Arabic texts into Latin from *c.* 1150 onwards which revived among western scholars the classical canons of scientific learning so long hidden from them. The Spain of the *reconquista* was the stage where this cultural exchange took place, and the most important individual was Gerard of Cremona, who worked in Toledo for almost half a century between 1140 and 1185 translating Aristotle and Ptolemy, and whose version of the *Almagest* spread Ptolemaic astronomy throughout Europe, providing the text of the first printed edition of 1515. Equally important were the translations of the *zijes* and manuals on the use of the astrolabe, which together placed in the hands of European astronomers the means to make their own observations and calculations. The most important tables

Above
Ottoman astronomers at work in the Istanbul Observatory, late sixteenth century.

were the 'Alfonsine Tables' promoted by King Alfonso the Wise of Leon and Castile and dated for the epoch of 1252, the eve of his coronation. They were disseminated throughout western Europe and helped to mould its astronomy for almost three centuries. By 1300 astrolabes were being made in Italy and France, and treatises on their use appeared in the vernacular languages, including Chaucer's version in 1391, written for his young son Lewis. One very specific legacy of Islamic astronomy was the naming of some fifty of the brightest stars. Greek sources had named only a handful of individual stars, such as Sirius, 'scorching', and Arcturus, 'bear-watcher', found in Homer and Hesiod. Most familiar star names, especially those beginning with 'Al-' (except Alcyone, which is Greek) are Arabic in origin: Algol, the 'demon' in Perseus, and Aldebaran, the 'follower' in Taurus, and so on. Many Arabic names were self-explanatory locations; for example, Mintaka in Orion means simply 'belt', and Markab, in Pegasus, 'shoulder'.

Science in the Christian West

The science of Christian Europe in the Middle Ages is the story of the encounter between conflicting intellectual authorities. The story cannot be written in terms of empirical or theoretical advances, but in terms of the evolving intellectual motives that controlled and shaped it. The dominating and unifying force in Europe in the centuries after the dissolution of secular Roman power was, of course, the Christian Church. Christian thinkers (like their Islamic counterparts) sought to construct a framework of learning about the creation as a whole with which their central religious beliefs would be in harmony. Cosmology became central, since an understanding of the mechanics of the universe would surely demonstrate God's rational power. Medieval cosmology was not monolithic; indeed it showed enormous ingenuity, and the idea that nothing happened in cosmology between Ptolemy and Copernicus is vastly oversimplified. But what did remain constant was an approach to cosmology that was determined by *a priori* beliefs, so that intellectual energy was poured into elaborate speculations in which philosophy, mechanics and theology were held in balance. Empirical research and mathematical modelling were at a discount.

The appeal which the closed, spherical universe had for the Greeks had been the appeal of geometric structures; to medieval Christian thinkers the appeal was theological, for all science was subservient to the religious motive. Questions concerning the structure of the cosmos, the shape of the world, the diversity of life, the origin of laws and society, were all seen as problems which related to the fact of God's rule over the universe and the human world. Deprived of the rationalising legacy of Greek thought, answers were sought in terms of *authority*; and the greatest authority of all was, of course, the Bible. Biblical texts which dealt with the divine ordering of the created world were examined and elevated into quasi-scientific dogma. Texts such as Job's 'Canst thou bind the sweet influences of Pleiades or loose the bands of Orion?'

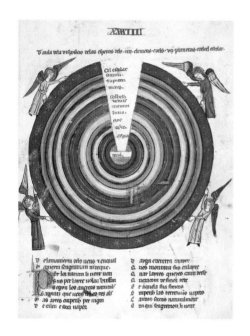

Above
The earthly elements and the paths of the planets are shown as circles on the cosmic sphere in this medieval cosmic diagram, the whole being upheld by four angels; the manuscript is 'Brevari d'Amor' by Matfre Ermengaud, French, early fourteenth century.

became the basis of an approach to cosmology that was neither rational nor empirical, but dogmatic.

Naturally it was recognised that the Bible had not pronounced on every conceivable subject, so recourse was made to secondary authorities, many of them late Roman authors such as Pliny, Martianus, Solinus and Macrobius, who offered encyclopedic but superficial collections of lore and opinion on philosophy, science and natural history. This form of literature was imitated by Christian encyclopaedists such as Isidore of Seville, whose works became in the middle ages an authoritative source on a whole spectrum of matters human and divine. In this milieu, science, philosophy and scholarship became overwhelmingly book-centred. The scholastic method of approaching any question was to ask: what have the authorities to say; how does Augustine, or Isidore or Bede treat this question? Neither reason nor experience were considered capable of outweighing the authorities of the past.

But there came a time when the church, the custodian of all learning whether sacred or secular, was shaken by the introduction of a new authority to challenge the old, namely reason. In the twelfth century, Abelard, Albertus Magnus and above all St. Thomas Aquinas evolved a rational approach to such questions as substance, volition, motion and causality which was quite new in the tradition of Christian philosophy. The source of this new method was the rediscovery of Aristotle's works, newly translated from Arabic into Latin, in which reason and experience were employed to answer questions of physics, cosmology, psychology or ethics. To Aquinas, theology was a science just as much as physics was: it was rational knowledge of God derived from irreproachable sources, namely revelation and reason. The mystery of God was expressible in human language, just as the mysteries of physics were, and therefore subject to the rules and structures of logical thought. Nature, Aristotle's *physis*, has its necessary laws, and the task of reason is to create a science, *logos*, of those laws. The crisis over Aristotelianism was not a doctrinal crisis, a question of new teaching about God or humanity, but a crisis of authority thrown up by this new approach: were the teachings of Christianity true because they were drawn from the Bible and the church guaranteed them, or because reason could approve them? Was the universe so because God willed it, or because it moved in obedience to rational laws of physics which even God could not alter, and which science claimed to discover?

Critical as this approach was for the church, it resulted directly in little new science, since the Nature that was lauded as the subject of rational science was in practice mediated through the theories and categories already devised by Aristotle. With the founding of the universities (Bologna, Paris and Oxford by the twelfth century; Cambridge, Padua and Rome by the thirteenth) the scholastic approach was, if anything, reinforced in the schools. Each discipline – mathematics, theology, law etc. – required fundamental teaching texts which were, of course, culled from the 'authorities'. In this context the Bible retained its pre-eminence, with its innumerable commentators, while Aristotle became the new and most sophisticated authority to be weighed against the more traditional ones. This intellectual

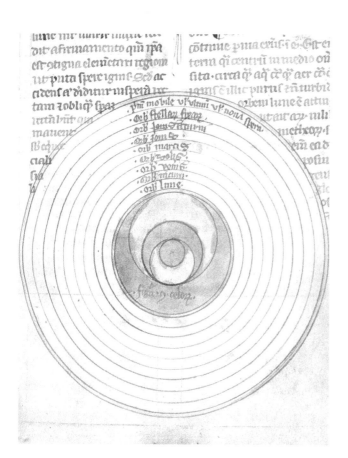

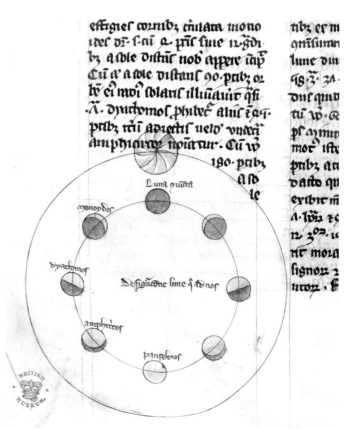

Above
The geocentric universe. The classical Ptolemaic structure of planetary spheres centred on the Earth, from a fourteenth-century manuscript of Sacrobosco's *De Sphaera*.

Above, right
Diagram from Sacrobosco's *De Sphaera*.

method clearly did not favour innovation in science. Astronomy is essentially an observational and deductive science, while the scholasticism of the medieval schools was overwhelmingly non-empirical. A question concerning celestial bodies and their movements was answered not by looking, measuring and deducing, but by referring to Ptolemy and Aristotle.

In the case of astronomy the most widely used text was *De Sphaera* by an Oxford scholar who taught in Paris, John of Holywood. He took the Latin name Sacrobosco. Clear, brief and entirely derived from classical sources, from c. 1230 onwards his text provided an accessible statement of the rudiments of Ptolemaic cosmology. The mapping of the heavens in the graphic sense did not exist, since there were no models for scholars to copy or elaborate. Many manuscripts of Sacrobosco's work were illustrated with small cosmic diagrams on the concentric-ring pattern. In the decorative art of the fourteenth and fifteenth centuries, it is interesting to see the vault of heaven filled with stars and having at its centre a circle or series of circles in which God is surrounded by the saints and angels. These images derive from the works of Dionysius, an obscure mystical Christian of possibly the sixth century AD, who wrote on the hierarchical orders of angels to be found in heaven. It was this cosmic structure which seemed to occupy the medieval mind rather than the relative locations of the stars. From the twelfth century there was a continuous awareness of Ptolemaic astronomy: manuscripts of the Latin *Almagest* were copied and circulated, complete with their co-ordinate tables and their workings in spherical geometry. Yet these two elements were never combined into maps

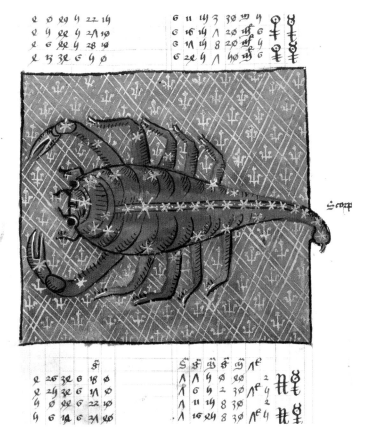

or globes. Nor did it occur to any scholars to transcribe into graphic form the northern star chart which they possessed in the rete of the astrolabe.

But alongside this inbuilt conservatism in western medieval science, there was a strong tendency towards the imaginative use of astronomy and cosmography. The ingenuity of the medieval mind, without which the great cathedrals could not have been built, produced a few technical advances of great importance, such as the mechanical clock, pioneered by Richard of Wallingford, Abbot of St Albans from 1327 to 1336. The clock provided philosophers with a satisfying new metaphorical image of nature, an image of a controlled mechanism which was to endure until the nineteenth century. The clock-face itself, which emerged in the fifteenth century, was clearly modelled on the face of the astrolabe, and its purpose was to symbolise the circular daily movement of the heaven through 360 degrees, mirrored in the movement of the clock hands through 360 degrees. At the same time a number of instruments had been devised to assist in astronomical calculations: the torquetum, the equatorium, the albion – these were all circular analogues of the heavens, as the astrolabe and the clock-face were, and as such they might be considered to be related to maps. Richard of Wallingford wrote advanced treatises on mathematics, and the image of medieval science as entirely abstract is sharply contradicted by the work of such exceptional figures.

Yet overwhelmingly the purpose of science and speculation was to understand nature as God's creation: all data and all theory must inevitably

Above
Richard of Wallingford with his clock, from a fifteenth-century manuscript. Richard, Abbot of St Albans, constructed one of the earliest mechanical clocks, which, in addition to keeping time, represented the movements of the Sun and Moon. Richard wrote advanced mathematical works, and devised the albion, a sophisticated scaled instrument for calculating planetary positions. In 1336 he died of leprosy, contracted during a visit to Avignon.

point to the end. For example, light was the subject of much study and speculation by Robert Grosseteste, Oxford's first chancellor. Light was considered on biblical authority to be the first creative principle, and a knowledge of its nature might lead to insights into the mind of God. Motion and causality were regarded as particularly important, for if everything that moves is moved, including the cosmos, all movement might be traced back to an ultimate source, which must be God.

This highly logical approach alarmed some churchmen, but its disciples attempted to disarm criticism with Aquinas's famous phrase that God was the author of 'the book of scripture and the book of nature'. Jean Buridan of the university of Paris meditated on the concept of impetus as a property which God imparted to the universe at its creation, as the clock is set in motion by its maker. Many contemporaries were apprehensive that such reasoning and speculation were fundamentally unorthodox, and in 1277 the church condemned a number of statements from the new rational theology. Some were abstract and hypothetical, for example that 'the first cause, God, could not make several worlds', because if he did they must be different from this, and this, as God's creation, must be perfect. This appeared to fly in the face of one of the basic tenets of Christianity, namely the omnipotence of God, and the fact that such speculations could acquire serious importance demonstrates the mutual interpenetration of science and theology. Yet medieval cosmology was far from monolithic, and even the fundamental Ptolemaic system was rejected by some. Ibn Rushd (known in the West as Averroës), the great Islamic philosopher, was an ardent Aristotelian, and could not accept the reality of the epicycles. According to Aristotle, circular orbital motion can only occur around a heavy body, and in space there are no heavy bodies: epicycles around nothing would be impossible. Averroës became widely known in the west and his doubts were carefully considered, but no alternative mathematical model to Ptolemy's could be found.

Yet the predominant motive of the new rationalists was not fragmentation but the search for harmony in nature and in thought. One of the strongest expressions of the medieval aspiration towards order was the work of Dante. The author of the *Divine Comedy* had a complete command of contemporary philosophy and science, and his constant use of them directs the reader to his sources – this was true for his contemporaries and remains true for us. The very structure of Dante's hell, purgatory and heaven are formed from a series of concentric circles, progressing ever towards a pivotal or controlling point. This schema was clearly suggested by the classical insistence that the circle or sphere was nature's most perfect form, and echoes the Ptolemaic system. That Dante was well aware of this system emerges clearly in the *Paradiso*, which is full of explicitly astronomical ideas; indeed, it is not too much to say that the central purpose of this final part of the poem is to offer a vision of the ordered harmony which God has built into the universe. Beatrice acts as Dante's tutor to explain the Moon's phases, eclipses and the motions of the planets, and it is no accident that the figure who guides them in the sphere of the Sun is the spirit of Aquinas himself. The miracle of Dante's freedom from earthly constraints, his journey through the spheres, echoes the deeply felt

Right
Astronomers using astrolabes. From
a fourteenth-century manuscript.

belief that the soul's true home was among the stars. The vision of a journey
through the cosmic spheres has sources in Plato and Cicero, and a later
echo in Marlowe's *Doctor Faustus*. Perhaps the central mystery which Dante is
exploring is how the mechanical universe can be moved by a spiritual force.
Dante seems to have felt that Aristotle's philosophy did provide an answer to
this metaphysical question through his definitions of matter and form. While
matter was the material substance from which things are made, form was
that which gave each thing its true quality. Form causes stars to shine and to
move, while in humans, form is the soul. Humans are, in fact, a conjunction,
a horizon, between soul and body, while the cosmos was imbued by God with
eternal harmonious motion, which expresses its form:

> ... Among themselves all things
> Have order, and from hence the form which makes
> The universe resemble God.
>   Dante, *Paradiso*, I, 100–102

Above
Medieval astronomer gazing at the stars. From a fifteenth-century French manuscript.

We cannot help noticing here that Dante's poem, and the various illustrations of heaven that appear in medieval manuscripts and paintings, all depict heaven as a real place in the sky, at some indeterminate distance above the Earth; but it is also a spiritual realm, the dwelling place of God and angels and of the souls of the departed saints. We should never underestimate the reality, as medieval thinkers saw it, of these two realms existing side by side: the realm of physical astronomy, and realm of spiritual forces. In studying and interpreting the physical heavens, the scientist-philosopher was approaching a superior spiritual dimension which was the home of God himself.

On a very different level from Dante's grappling with metaphysical forms, other medieval writers made a more transparent use of astronomical symbolism. By Chaucer's time the astrolabe had arrived in western Europe along with much astrological lore translated from the Arabic. It has only recently been understood how extensively Chaucer's writings embody the spirit of the age by weaving astrological concepts into the structure of his plots. Built into the narrative of *Troilus and Criseyde* is a pattern of planetary conjunctions which counterpoint the rise and fall of human ambitions in love and war. Even more specific are some of the *Canterbury Tales* such as 'The Nun's Priest's Tale', which has an astrological subtext. The story concerns a farmyard cock, his wives and his narrow escape from a fox. These characters correspond to celestial bodies – the Sun, the Pleiades and the planet Saturn respectively – and the story is related to four different arrangements of the heavens during the day of the drama. Since Chaucer was an expert on the use of the astrolabe, some scholars have suggested that he may have demonstrated this hidden meaning with the astrolabe after reading the poem. Later, after the advent of printing, paper astrolabes were commonly printed in astronomical books, to be cut out and mounted for practical use.

Chaucer died in 1400, on the eve of the new century in which celestial mapping finally emerged in Christian Europe, partly from Arabic sources, but also as a feature of the nascent scientific humanism associated with figures such as Nicholas of Cusa, Peuerbach and Regiomontanus. Nicholas of Cusa, a cardinal of the church, was a wide-ranging scholar some of whose speculations have led to his later reputation as a prophetic figure, even a precursor of Copernicus. Certainly he drew attention to the fact of relative motion, and suggested that it was possible that the Earth could move just as the other objects in the cosmos did. He took a keen interest in practical science, and collected many instruments, including the earliest known western celestial globe. Georg von Peuerbach and Regiomontanus (real name Johann Müller, whose home town of Königsberg provided his Latin name) developed the study of astronomy in Vienna and later in Nuremberg, where they made an edited version of Ptolemy's *Almagest* and new astronomical tables. These works achieved special fame for their authors as the first printed works on astronomy.

The advent of the new medium meant that astronomical data and theories could reach a wider audience, and the simultaneous first printing of Ptolemy's work on terrestrial geography drew attention to the possibilities of using co-ordinate systems and map projections in the creation of paper maps. Many editions of Ptolemy's *Geographia* contained the first scientifically constructed maps ever seen in Europe, and it was only a matter of time before star charts constructed from the data in the *Almagest* would be drawn and printed. It was at this period too, *c.* 1470–1520, that the great voyages of exploration first took European seamen far beyond the familiar waters of the Mediterranean. The problems of navigation this presented would not be solved with new terrestrial maps for a further century or more. But innovative mariners and technically minded mathematicians were stimulated to devise new forms of instruments to facilitate navigation, and some form of two-dimensional star chart and three-dimensional star globe were basic requirements. Regiomontanus worked from 1467 to 1471 in Hungary under the patronage of the humanist monarch Matthias I, among a circle of scholars which included Martin Bylica and Hans Dorn, who made and used the first celestial globe known to have been made in the west, a very fine instrument very much on the Islamic model.

Astrology

Knowledge of astrology in any detail had been virtually extinct in the west for many centuries until the process of translating classic Arabic treatises on the subject, such as Abu Mashar's, led to its revival. It is often said that in pre-scientific ages astrology and astronomy were the same thing. This is only true insofar as mature astrology is based on the computation of celestial positions; therefore a knowledge of the rules and language of astronomy was essential. But it was perfectly possible for a late medieval scholar to study only the mathematical, physical side of astronomy, just as it was equally possible for the astrologer to devote himself exclusively to its philosophical side, accepting his data from others. An example of the first school would be Nicholas Oresme (1320–82), the highly original French mathematician who likened the cosmos to the newly invented mechanical clock, and who speculated on the possibility that the Earth rotated in space. Yet he still spoke of the spheres being moved by intelligences and neither he nor any contemporaries sought to deny that the stars and planets do indeed influence human life. This influence, he thought, was a question of nature, deriving from the qualities of those bodies, which might be studied and classified. On the other hand the attempt to predict the future, to manipulate fortune, savoured of magic: it compromised human freedom and should be shunned. The danger of fatalism was one that haunted many Christian and Islamic thinkers. Yet there were also secularly minded thinkers who were impatient of the great tortuous labyrinth of astrology. Ibn Khaldun, the great fourteenth-century Arab historian, could write that 'Astrology is all guesswork and conjecture, based on the assumed existence of astral influence, and a resulting conditioning of the air'.

These were the views of an exceptional intellect, and they were undoubtedly running against the tide. The medieval conviction that

hierarchical order permeated the cosmos led thinkers to seek for links in the 'chain of being' that bound man and nature together. In *Tetrabiblos*, Ptolemy had suggested that a power (*dynamis*) radiated through the ether, causing changes in all the four earthly elements found in humans, animals and plants. The Moon, for example, affected human health because it caused a tidal flux in the body's humours. Given the keys to interpret these effects, it should be possible to predict in advance both natural events and the course of human life. Albumazar had argued that the planets were deliberate dispensers of forces, that they were intelligences, and since their effects could be rationalised, they must be rational. This belief was further strengthened by the fact that they were apparently self-moving, unlike the fixed stars, and must therefore be possessed of will and purpose. Much later, the more scientific Robert Grosseteste even spoke of the celestial influence falling on the Earth like rays of light in complex geometric patterns, and he suggested that it was the variations in these patterns that affected human character and events. But if the planets were intelligences, how were they bound to their planetary bodies, and what was the nature of those bodies? It was to questions like these that medieval scholars devoted a good deal of imagination. Dante even speculated that ideas might pass to us from the stars. This universal belief in celestial influence led to a view of the heavenly bodies as intermediaries which God employed to shape human life. On this view, astrology could be presented as compatible with orthodox religious faith – it was yet one more statement of man's organic link to the rest of creation.

This belief was elaborated in a vast body of accepted lore. Carnal humanity was linked to the animals, plants and minerals of this Earth, and a knowledge of their properties was essential to human bodily health. But the human soul was linked to the higher levels of creation: the Moon, stars and planets. One of each of the four earthly elements – earth, air, fire and water – dominated in each person and helped to determine his or her character. Each of the planets which might reign at the time of nativity produced characteristic types: Jovial, Martial, Saturnine etc. The planets and the Zodiac signs influenced both body and mind, and thus the science of medicine in the Middle Ages was permeated by astrology. Not only individuals but professions, cities and nations had their stellar and planetary signs: for example thieves, shopkeepers (and later printers) were under Mercury, artists and musicians under Venus. Virgo was the star sign of Paris and Heidelberg, while Sagittarius ruled Cologne and Avignon. It is noticeable that while star charts are unknown at this period, emblematic figures of the separate constellations and planets appear prolifically in manuscript illustrations, modelled initially on those of Al-Sufi.

These levels of popular belief were vastly removed from the philosophical subtleties of Aquinas or Dante, but astrological emblems certainly flourished alongside the saints and the other Christian imagery. Judicial astrology was that branch of the art which sought to identify the most propitious moment for an enterprise – a marriage, a battle, a coronation, and so on. Thus political and courtly life provided the social setting in which, among coteries

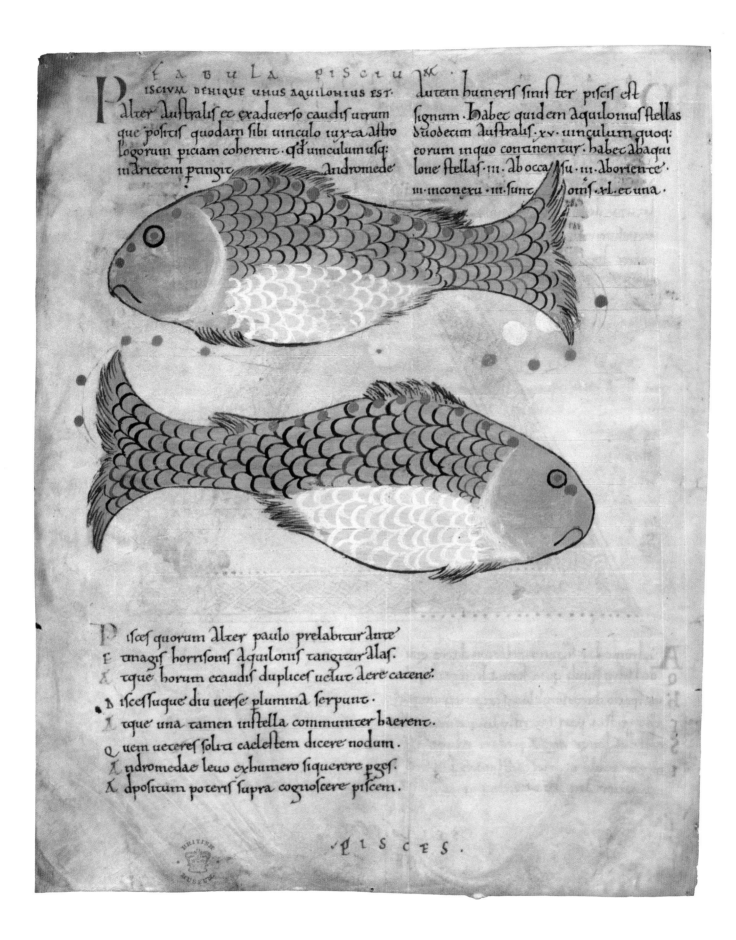

FABULA PISCIUM

Piscium denique unus aquilonius est.
alter australis et exaduerso caudis utrum
que positis quodam sibi uinculo iuxta astro
logorum piciam coherent. qd uinculum usq
in arietem ptingit.   Andromede

autem humeris sinister piscis est
signum. Habet quidem aquilonius stellas
duodecim australis. xv. uinculum quoq
eorum inquo continentur. habet abaqui
lone stellas .iii. ab occasu .iii. aboriente
iii. inconexu .iii. sunt   ....onis .xl. et una.

Pisces quorum alter paulo prelabitur ante
Et magis horrisoni aquilonis tangitur alas.
Atque horum ecaudis duplices uelut dere catene:
Discessuque diu uerse plumina serpunt.
Atque una tamen instella communiter haerent.
Quem ueteres solita caelestem dicere nodum.
Andromedae leuo exhumero siquerere pges.
Apositum poteris supra cognoscere piscem.

PISCES.

Above

Christianity and Astrology combine in this 15th-century Book of Hours God as creator is juxtaposed with Jupiter, his pagan equivalent, and with Aries, the first sign of the zodiacal year. The persistence of these pagan mythological emblems and themes throughout the Christian middle ages demonstrates the unshakeable hold which astrology maintained on the popular mind.

and intrigues, astrology flourished, and where the court astrologer was often the court doctor. The feverish climate of the fourteenth century – stirred by the Black Death, the Great Schism in the church, long wars between England and France, the radicalism of the Lollards and rumours of the Antichrist – all served to turn the mind to divination, and to lead scholars to seek in the stars the fate of the principalities and powers whose future appeared to be so troubled. Of course the actual predictions of the astrologer rarely came true, but far from weakening the art this had the effect of increasing the volume and complexity of astral lore, in a search for remedies for man's defective knowledge. That astrology was respectable and not automatically in conflict with the Christian religion may be gauged from the appearance of zodiacs in many medieval churches. Of course a Zodiac is not purely astrological; it can be a symbol of time, of the cycle of the year, and in this sense its presence in a church may be simply a reminder that God was the lord of time, and of the heavens. But there are certain Zodiacs which are given such a prominent place in a church that it seems clear that a pro-astrological statement was being made; a good example is the large elaborate Zodiac carved in coloured marble on the floor of the nave in the Basilica of San Miniato al Monte in Florence.

Astrology was crucial in providing the intellectual framework for another science, namely medicine. Without exception, all the authorities in the western Middle Ages, in the Islamic world and in the Renaissance period accepted that astrological learning was central to medicine. As one of them put it, 'All things here below – air, water, the complexions, sickness and so on – suffer change in accordance with the motions of the planets.' The four elements of classical and medieval science, earth, air, fire and water, manifested themselves in the human body as the four humours, and these humours were in turn responsible for the four temperaments: sanguine, melancholic, phlegmatic and choleric. Every human individual was a blend of all four qualities, but with one usually predominating. When the balance of the humours was disturbed, sickness resulted.

The greatest single cause of any such imbalance was believed to be the movements of the planets, and especially their positioning within the signs of the Zodiac, each sign having a special power over a particular part of the body, from Aries at the head to Pisces at the feet. This scheme had been arrived at by outspreading the image of the human figure upon the zodiac circle, thus giving forceful expression to the supposed correspondence between man and the cosmos. The physician would invariably begin his treatment by casting the patient's horoscope, to find out first his humour, then to analyse the heavens at the moment when the sickness began, and finally to decide on a propitious moment to treat the malady. Medicine on these principles was taught in all the universities of Europe from the twelfth century onwards, where it took its place beside philosophy, mathematics and astronomy, all these subjects being based on classical sources as mediated through Islamic scholars. Not only the health of the individual was analysed in this way, but the health of whole societies – indeed, the world. During and after the great outbreak of the Black Death in 1347, which may have killed a third of the population of Europe, the plague was widely believed to be the result of a rare and fatal conjunction of the planets Saturn, Jupiter and Mars in the sign of Aquarius , causing a great and terrible 'corruption of the air', a judgement that was endorsed by the University of Paris. Episodes such as this show that astrology had taken its place at the centre of intellectual life in Europe, and it would be many centuries before that place was challenged.

The heady marriage of astronomy and occult science did not suddenly dissolve in the Renaissance period, which we see as the end of the Middle Ages, but it did become distinctly less Christian and more overtly pagan. The Renaissance largely reshaped European art, politics and religion, but science lagged behind. Instead, the sixteenth century witnessed a luxuriant growth of astrology as part of a fashionable cult of nature mysticism. In particular, the Renaissance admiration for all things classical led to the re-discovery of a body of supposedly ancient Greek writings on magic, alchemy, astrology and other forms of occult learning, which were brought to Florence from Greece. The name attached to these manuscripts was that of Hermes Trismegistus ('Thrice-great Hermes'), the Greek equivalent of Thoth, the Egyptian god

Zodiac Man from a fifteenth-century
English manuscript. One of the
central doctrines of astrology was
that the parts of the human body
were influenced by the planets
and by the twelve signs of the
Zodiac – Aries ruled the head,
Pisces the feet, and so on – and
that the planets' presence within
the star groups produced malign or
beneficent effects. The movable dial,
or volvelle, enabled the physician
to calculate planetary positions
and complete his diagnosis.

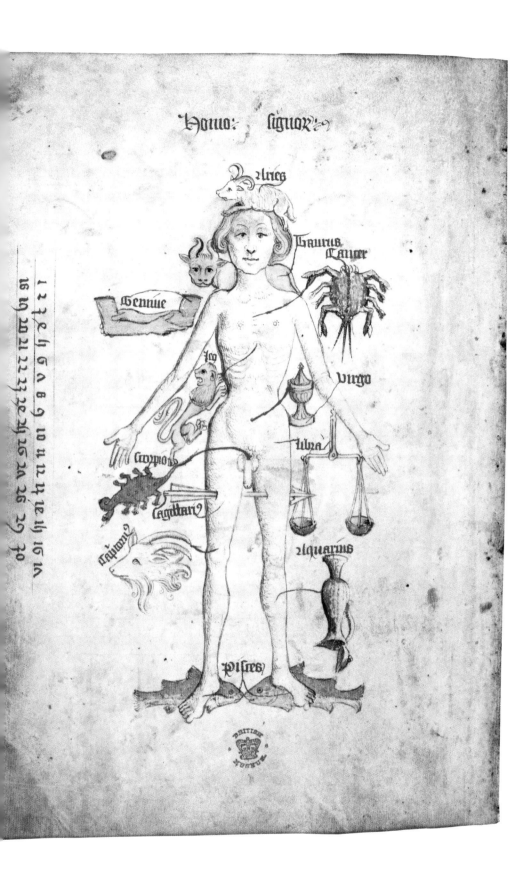

of wisdom, language and magic. They purported to be older than the delivery of God's law to Moses, and in some ways they presented an alternative vision of God and humankind. They taught that humans potentially had the power to unlock nature's secrets, to manipulate invisible forces in the Earth, in the sky and in the living world, through magic formulae and ceremonies. Their importance for astronomy/astrology was their teaching that humans could draw down the hidden, unknown powers of the stars and planets and bend them to their own will. The translator of these texts, the Florentine Marsilio Ficino, even suggested that Hermes might have been Moses under another name, giving to the world a divine but secret revelation. The great fashionable appeal of Hermetic philosophy can be seen in the portrait of Hermes inlaid in marble on the floor of Siena Cathedral, an extraordinary tribute to this mythical figure, who is there hailed as a prophet of Christ.

In fact the Hermetic writings had been known centuries earlier to the church Fathers, who had attacked them as heretical, but they had been lost sight of. When these documents had been rediscovered and translated into Latin, their ideas were received with intense excitement by a new generation of occult philosophers such as Cornelius Agrippa, Tomasso Campanella, Giordano Bruno and Giovanni Pico della Mirandola. Agrippa is sometimes thought to have been the model for Faust, the archetypal Renaissance magician who gave his soul in exchange for forbidden, diabolical powers. These figures were striving to break out of the sterile world of orthodox religion and academic philosophy, and into a more powerful and personal understanding of nature, in which humanity might elevate itself to the level of the gods. One of Faust's first demands, having concluded his pact with the devil, was to be taken on a journey through the heavenly spheres and shown the secrets of the cosmic structure. In a strange way, this ambition to seize secret powers and transform a mortal into a god was prophetic of what technological humanity would indeed become several centuries hence, with the scientific revolution and the industrial revolution. The rhetoric of Renaissance occultism heralded in an uncanny fashion the Sun-centred theory of Copernicus. Marsilio Ficino, the earliest exponent of the Hermetic philosophy, could write:

> Therefore if you wish to see God, consider the Sun, consider the path of the Moon, consider the order of the stars. Who is it that keeps this order? The Sun, the highest God among the gods of the heavens, to whom all the other celestial gods give way as to a king and a master ...

But the power and attraction of the Hermetic philosophy came crashing down in 1614, when the great classical scholar, Isaac Casaubon, proved that both the language and the contents of the texts must be dated to the second century AD, and therefore did not preserve a pre-Christian or pre-Mosaic revelation. So one potential avenue of development for Renaissance science was closed off, to be replaced by another very different one. Certain implications of Copernican theory – the centrality of the Sun and the newly discovered vastness of the cosmos – had suited the Renaissance occultists perfectly, while orthodox religion retreated in fear. The fate of Giordano Bruno, burned in Rome for his blend of Copernican and occult Hermetic beliefs, demonstrated the still indissoluble link between astronomy and the foundation of faith. In this period before the scientific revolution, astronomy existed partly in a dimension of high-flown speculation, just as Hermetic magic had, in contrast to the relative scarcity of observational work. The reversal of this balance in the coming century was to revolutionise astronomy.

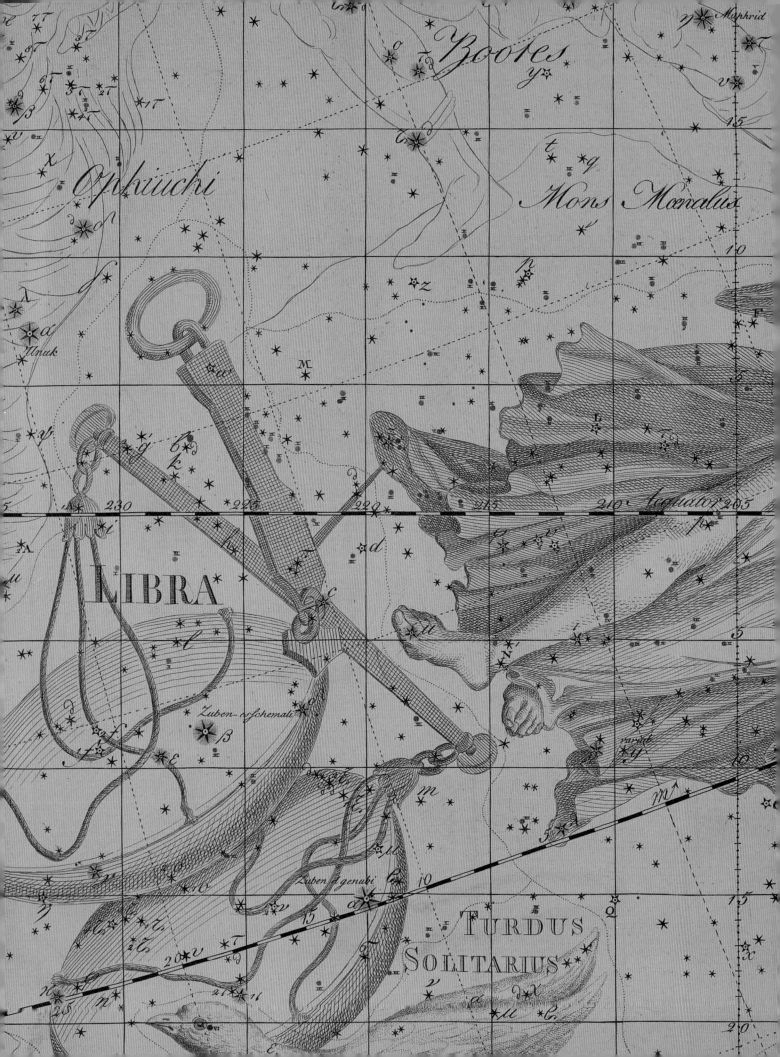

Maphrid

Bootes

Ophiuchi

Mons Mænalus

Unuk

LIBRA

230   225   220   215   210 Aequator 205

Zuben-eschemali

Zuben el genubi

TURDUS

SOLITARIUS

CHAPTER THREE

The New Science

'Two things fill the mind with ever new and increasing admiration and awe, the oftener and more steadily they are reflected on: the starry heavens above me and the moral law within me.'

Kant, *Critique of Practical Reason*, 1788

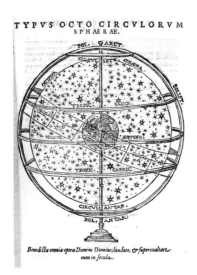

TYPVS OCTO CIRCVLORVM
SPHAERAE.

Benedicta omnia opera Domini Domino laudate, & superexaltate eum in secula.

Above
Giovanni Gallucci's armillary sphere,
1588. The sixteenth century saw an
intense interest in scientific pictures
and diagrams. Here the traditional
outline armillary is given a naturalistic
surface of five- and six-pointed stars,
although the draughtsman has not
formed them into constellations.

Opposite
Frontispiece to Galileo's *Dialogus
de Systemate Mundi*, with images
of Aristotle, Ptolemy holding an
armillary sphere, and Copernicus.

For the centuries so far covered by this book, *c.* 3000 BC–AD 1500, just as
terrestrial cartography in our sense scarcely existed, so celestial mapmaking
did not exist either. There were moments during the long era before the age
of modern western science when it might have emerged. One moment must
surely have been with whoever designed and constructed the Antikythera
mechanism, probably around 100 BC, for which many preparatory drawings
and models must have been made to analyse the structure of the heavens.
A second opportunity came with the work of Ptolemy (*c.* AD 150) who
designed, and probably made, a celestial globe like the marble one we see
carried by the Farnese Atlas. But he did not take the step of transferring that
concept to two-dimensional maps. Intriguingly, this was also the case with
his terrestrial mapping: he set out in detail a theoretical basis upon which
to construct maps of the known world, but there is no evidence that he
actually drew them. Another potential breakthrough lay in the hands of the
Islamic astronomers of the tenth to the twelfth centuries, who produced the
astrolabe, but who once again did not take the step of transferring to paper
the map which it contained. Presumably the sophistication and flexibility of
the astrolabe meant that they had no reason to do so.

Throughout its history astronomy had its technical basis in mathematics,
linear or geometric. That celestial mapping did not emerge earlier is more a
matter of cultural and social context than fundamental intellectual obstacles.
In none of the cultures in which astronomy flourished was there a tradition
of scientific diagrams, or of visualising philosophical concepts. The absence of
a cartographic language is but one symptom of this. This situation changed
emphatically in Renaissance Europe, where the natural world became the
subject of visual interpretation through experimentation and study in physics,
mechanics, optics, dissection and conceptual modelling. Celestial mapping
only emerges in the late fifteenth century, in parallel with the development
of terrestrial mapping. Matters such as projections and co-ordinates had
certainly formed part of Greek science, and were dealt with extensively by
Ptolemy; but for all practical purposes these techniques were dormant during
the Middle Ages, and re-emerged in the late fifteenth century in the context
of the revival of Ptolemaic mapping. The result was a new science of ordered
space, of measuring the world and also representing it in mathematically
controlled diagrams, for example using the co-ordinate system. This new sense
of ordered space coincided with the age of printing, and, somewhat later, with
the age of the new empirical astronomy associated with accurate sighting
instruments. The theories and discoveries of Copernicus, Tycho, Kepler and
Galileo stimulated an intense and widespread interest in astronomy.

It is no accident that calendar reform, overdue in Europe for many
centuries, was finally enacted in the later sixteenth century. The Julian

DIALOGVS
DE SYSTEMATE MVNDI,
Autore
GALILÆO GALILÆI LYNCEO,
SERENISSIMO
FERDINANDO II. HETRVR. MAGNO-DVCI
dicatus.

ARISTOT.   CL. PTOLEM.   N. COPERNICVS.

Augustæ Treboc.
Impensis BONAVENTVRÆ et ABRAHAMI ELZEVIR
Bibliopolar. Leydens.

calendar year of 365.25 days was too long, and produced an error of eleven minutes per year. One thousand years after its introduction in the first century BC this had accumulated to seven days, and by the sixteenth century the vernal equinox, used to determine the date of Easter, had moved ten calendar days from its true date, a situation intolerable both to churchmen and to scientists. The complexities of the problem were great, however, and of the many leading astronomers who were consulted, some (like Copernicus) refused to pronounce on the subject, and many years of thought, discussion and calculation were needed before the new calendar was instituted by Pope Gregory XIII in 1582.

In response to the new astronomical awareness of the late sixteenth century, the celestial map, as a conceptual model of the heavens that was easy to produce and to handle, was offered by map-publishers to an increasingly literate and scientifically minded population. The structure of the star chart, the projection of a measured sphere, was dependent on the new language of cartography which appeared at the end of the fifteenth century. The key features of that language, without which modern scientific mapping could not emerge, were the science of map projection and the co-ordinate structure, the ordered space imposed by the grid of latitude and longitude, that was learned by Renaissance geographers from the revived works of Ptolemy. The new art of 'cosmography', with its diagrams of the Earth and the heavens, became a characteristic Renaissance pursuit. With the growth of map printing, most atlases from the later sixteenth century onwards included star charts of the northern and southern heavens, often with diagrams of cosmic structures, the geometry of eclipses, lunar phases and so on. The star chart, a scientific document just as the world map is, became a publishing genre, subject to the intellectual and commercial demands of the day.

The Copernican Revolution and its Aftermath

The sixteenth century witnessed a quickening of scientific thought and the virtual rebirth of astronomy in its modern form, driven by a new approach to the problem of cosmic structure. The awakening of science came noticeably later than the revival of art, scholarship, discovery and commerce which we call the Renaissance, and when it did come in the later sixteenth century, the intellectual focus, as so often before in astronomy, was on the structure of the universe, the relation between the Earth, Sun, planets and stars. The Copernican Revolution in astronomy which placed the Sun at the centre of the solar system has been widely and rightly seen as one of the great intellectual turning-points in human history, causing a profound shift in humanity's understanding of the universe. With the perspective of history this is undoubtedly true, but it must be emphasised that this was no sudden cataclysm: the Copernican model won acceptance very slowly, and, almost a century after his theory was given to the world, serious astronomers could still be found who rejected it entirely. The full implications of his revolution emerged only towards the end of the seventeenth century in the work of

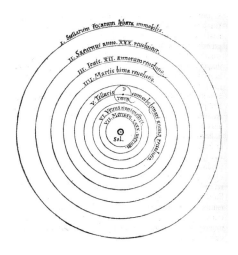

Above
The Copernican system: the astronomer's own diagram.

Newton, while the intervening century and a half had been a period of deep uncertainty, hypothesis and experiment. It is well known that Copernicus himself hesitated for many years before placing his theories before the world, almost certainly because he foresaw the controversy they would cause.

On the evidence of an essay which he circulated in manuscript, his ideas were essentially formed by 1510, while according to tradition a copy of his printed book *De Revolutionibus Orbium Coelestium* was placed in his hands on the very day of his death in 1543. There is no record that any one great insight or inspiration led Copernicus to his new theory. He was aware of the tradition that certain Greek thinkers such as Aristarchus of Samos (active *c.* 300 BC) had suggested that the observed movements of the heavenly bodies might be explained very simply by the motion of the Earth itself, a motion that was however ruled to be impossible by Aristotelian physics. But having once seriously investigated this 'impossible' idea, Copernicus realised how elegantly a heliocentric theory explained the most puzzling type of planetary motion – the retrograde path. Between 1510 and 1540, he accumulated observational data and worked out detailed models of the paths of the planets around the Sun. The result was a graceful, geometrically satisfying system, but one which was powerless to explain the physics that lay behind it: since it was clearly not fixed on a crystal sphere, how could the Earth move? Copernicus's meditation on a Sun-centred cosmos was not purely cerebral and geometrical. There is some evidence that, perhaps during his youthful studies in Italy, he had become aware of the Hermetic philosophy, in which fire and the Sun were creative and purifying forces. This mystical language finds a strong echo in Copernicus's rhetorical description of the Sun as 'the lamp in the temple', 'the soul of the world', and 'a visible god'.

Of the many profound implications that flowed from Copernicus's work, two of the most immediate were the size of the universe and the problem of falling bodies. First, if the Earth were truly circling the Sun at a distance which Copernicus estimated (wrongly) to be some 4.5 million miles, there should be quite clear stellar parallax: the stars on the starry sphere should move dramatically in relation to each other. Of course they did not, and the inescapable conclusion was that the stars were so distant that no parallax could be observed. While Copernicus did not overtly question the reality of the heavenly spheres – indeed he retains the word in his book's title – a universe of unsuspected vastness began to reveal itself. Secondly, classical and medieval physics affirmed with Aristotle that objects on Earth fell in natural motion towards the still centre of the universe. But if the Earth were neither still nor central in the universe, a new theory of weight and perhaps of motion was required. Copernicus's theory plainly raised far-reaching questions which in itself it could not answer, and some of his scientific contemporaries took refuge in the fiction that Copernicus had devised an abstract geometric model not intended to represent physical reality. A succession of outstanding astronomers set themselves the task of bridging the gulf between geometry and physics, a task in which empirical observation was seen to be crucial.

The first was Tycho Brahe, whose precise observations of the paths of comets across the solar system in the 1570s drove him to suspect that the

103   The New Science

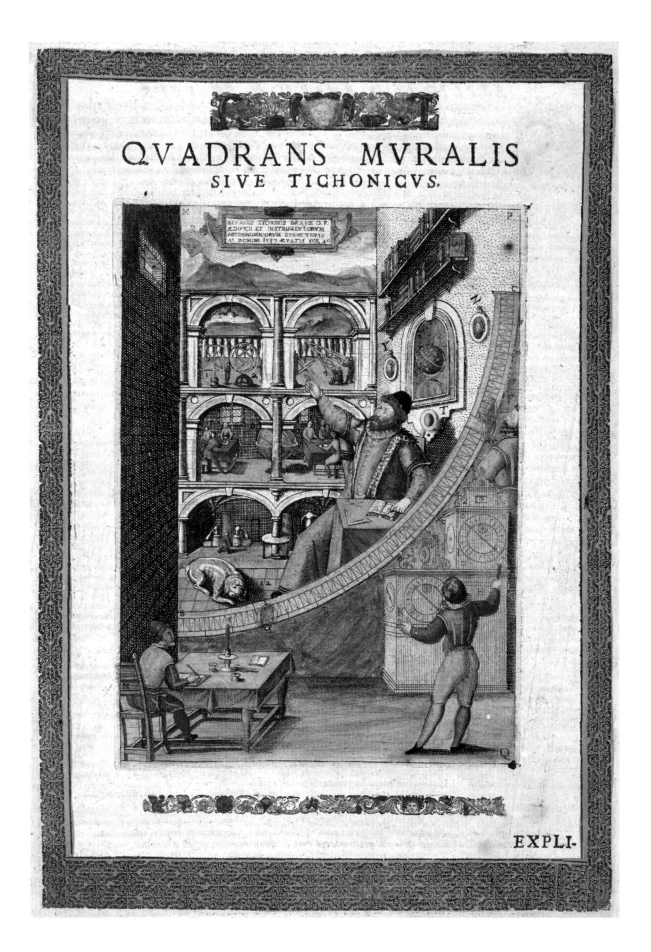

EXPLI-

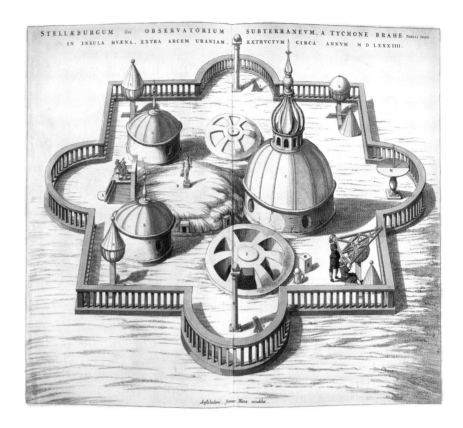

Opposite
Tycho Brahe, 1598. A famous picture of the great astronomer in his observatory. Tycho's work was pre-telescopic, and the large mural quadrant with which he observed all his star positions is seen in the foreground.

Right
Tycho Brahe's observatory. Plate from Blaeu's *Atlas*.

crystal spheres were a myth. He also studied the appearance of a new star – a supernova – whose brief, striking and mysterious career in 1572–73 plainly challenged the classical doctrine that the celestial realms were unchanging and incorruptible. Tycho set new standards of observational thoroughness at his celebrated observatory Uraniborg ('Castle of the Heavens') on the Danish island of Hven, yet he was one of those who could not reconcile himself to the Copernican system. A moving Earth was an absurdity to him, and he devised a compromise theory in which the five planets did orbit the Sun, while the Sun still circled the Earth. Yet this required the paths of planets to cross each other, which would have been impossible in the presence of the celestial spheres. This theory was published in 1588 and it attracted much support because it confirmed the evidence of our senses that the Earth does not move, and because it re-enthroned humanity at the centre of the universe. Tycho constructed a large, elaborate celestial globe which must be counted one of the most important in the history of astronomy. It has not survived, but he left a description of it in his great book on astronomical instruments published in 1598. It was almost 5 feet in diameter, formed of wood covered with thin brass sheets on which were etched the ecliptic, the equator and a meridian. The calibration was of unmatched precision, with each minute of arc being engraved. As his observation and cataloguing proceeded, Tycho marked each star on this globe, so that by 1595 it displayed the locations of 1,000 stars.

It is easy to use the phrase 'new star catalogue', but this accurate positioning of hundreds of individual stars depends on the meticulous

measurement on an invisible co-ordinate system of a few fundamental
stars to which the others may then be related. Tycho laboured for years to
secure instruments of unprecedented accuracy: they were all pre-telescopic
of course, and consisted of various large sighting devices such as the famous
mural quadrant and the huge armillary. Tycho's was a radical commitment to
a new type of empirical astronomy: if the astronomer wished to understand
the heavens he must observe and measure them, he must see them as they
are, not theorise or be enslaved by dogma. The fact that Tycho produced no
star maps perhaps tells us something of their limited role. His star catalogue
was for the committed professional astronomer who would use it with his
own instruments to observe the stars. A map was a demonstrational aid

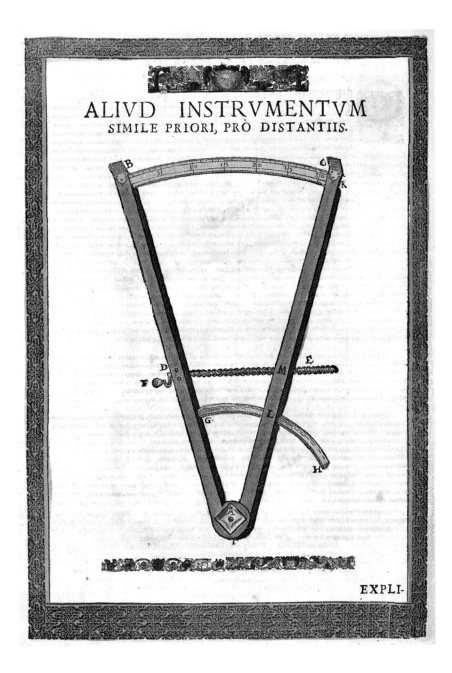

which the non-specialist without instruments might use to identify what he saw in the sky. The distinction is an important one, for it explains why the star chart became the province of the commercial map publisher, while for the serious astronomer it was the star catalogue which mattered.

The astronomer whose achievement was to bridge the gulf between theoretical geometry and physical motion was Johannes Kepler. He worked within the Copernican framework, and from his earliest career his thought tended to the philosophical, to the quest for ultimate causes in the universe. He was a creative astrologer, committed to the search for the links and harmonious relationships among the many realms of nature, including the human. He was convinced that the soul received at birth the imprint of the

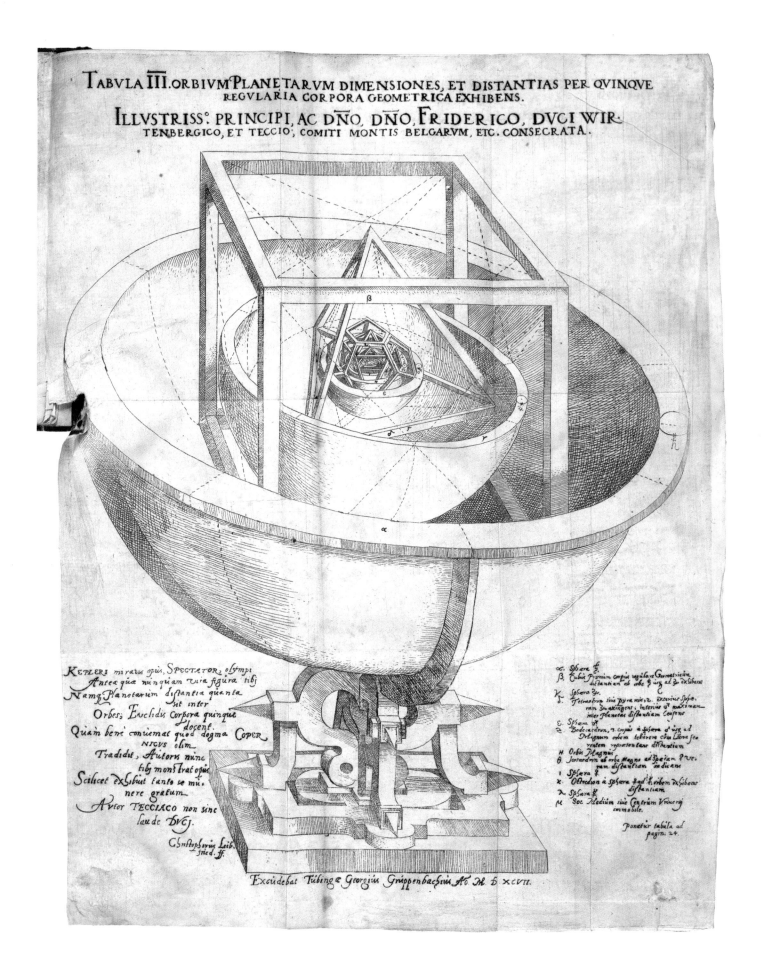

TABVLA III. ORBIVM PLANETARVM DIMENSIONES, ET DISTANTIAS PER QVINQVE REGVLARIA CORPORA GEOMETRICA EXHIBENS.

ILLVSTRISS: PRINCIPI, AC DÑO, DÑO, FRIDERICO, DVCI WIRTENBERGICO, ET TECCIO; COMITI MONTIS BELGARVM, ETC. CONSECRATA.

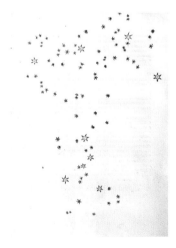

Above

Galileo: Map of stars in Orion's belt and sword. The first star map made with the aid of a telescope: Galileo found some eighty stars, where the naked eye sees perhaps a dozen.

reigning celestial pattern, and he pondered the possible mechanisms by which this might operate. For a time he considered that the quality of light radiating from each celestial body might vary, and that light could therefore be the medium of cosmic force. He also explored the correspondences between celestial configurations and meteorology, in the tradition of Ptolemy's *Tetrabiblos*.

One of Kepler's most famous attempts to lay bare the secret harmony of the universe was his geometric model of the solar system. Apparently he discovered in a flash of inspiration that the ratios of the planetary orbits conformed to a series of regular solids: between Saturn and Jupiter the cube, between Jupiter and Mars the tetrahedron, between Mars and Earth the dodecahedron, between Earth and Venus the icosahedron, and between Venus and Mercury the octahedron. These are the five regular solids of classical geometry; there are no others. This intriguing model does indeed approximate to the truth, but it transpires to be an isolated relationship, unrelated to any laws of science. Kepler saw it as confirming his sense of cosmic harmony, and he continued his researches into the planetary paths. Between 1609 and 1619 he published what became known as his three laws of planetary motion: that the orbits of the planets are not circles but ellipses with the Sun at one focus; that their radius vectors sweep out equal areas in equal time, hence in the elliptical orbits their velocities cannot be constant; and that their orbital period is in direct mathematical relationship to their solar distance, that is, the cube of the distance is in constant ratio to the square of the time (e.g. Jupiter is 5.2 astronomical units from the Sun; 5.2 cubed is 140.6, and the square root of 140.6 is 11.9, which is Jupiter's orbital period in Earth years). The significance of these discoveries was immense for they signalled the approaching end of Aristotelian physics, making possible their replacement by empirically derived laws of mechanics. Abstract doctrines such as the necessity of pure, spherical motion were banished, as was the complex network of epicycles, which had survived both Copernicus and Tycho. Kepler had not, perhaps, discovered the underlying cosmic harmony which he sought, but his dynamic model was an account of physical reality, not an elegant hypothesis. Yet the motive power of the model, the forces which held the universe in equilibrium, still remained as mysterious as ever.

The seal was set on this remarkable period of astronomical innovation by the discoveries of Galileo. Less profound as a theorist than Copernicus or Kepler, Galileo used the newly invented telescope to look with critical eyes at the solar system, and he rigorously re-interpreted what he saw. When he trained his telescope on any region of the sky, apparently empty space was resolved into previously unseen star-fields. He illustrated this by publishing a new map of part of the constellation Orion, showing some eighty stars, where the naked eye could detect no more than a dozen. The concept of scale cannot be precisely applied to celestial maps, but Galileo's achievement here may be thought of as the equivalent of producing the first large-scale map of one of the constellations. None of his discoveries was more significant than that of the moons of Jupiter, announced in his famous book *Sidereus Nuncius*

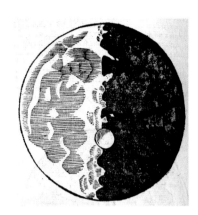

('The Starry Messenger') of 1610. Here was a system of celestial primary-and-satellite exactly mirroring the supposedly unique relationship of the Sun and its planets. Clearly there were many primary centres of motion in the solar system, and therefore laws of motion which were valid throughout the universe. This discovery now undermined the former view of the uniqueness and centrality of the Earth. Galileo also laid claim to discovering the first empirical proof of the Copernican theory of a Sun-centred system. This proof was the phases of Venus, resembling those of the Moon: he argued that Venus would exhibit these phases only if it were orbiting the Sun, whereas in the Ptolemaic system Venus always lies between the Earth and the Sun, and therefore could never show a fully illuminated face. The phases of Venus are not visible to the naked eye, but they were visible even to Galileo's primitive low-power telescope. With his pioneering telescopic work, Galileo's achievement was to show that the heavenly bodies were simply not what people had always believed them to be, and he revealed these discoveries in a series of vivid printed images.

Contemporary with Galileo, one of the foremost philosophical minds of the age turned his attention to the baffling problems of cosmic mechanics thrown up by the new science. René Descartes proposed to replace the celestial spheres by vortices of invisible, subtle matter, a kind of cosmic streaming, which carried the Earth and the heavenly bodies in their paths. It was as ingenious as it was unverifiable, and the wide support it attracted throughout the seventeenth century shows clearly the extreme difficulty that scientists and philosophers had in conceiving that the universe could be held together without a physical framework: the concept of empty space was unthinkable. The addiction to the idea of direct physical cohesion that lay behind the classical notion of the spheres, and behind the Cartesian vortices, was finally vanquished by the genius of Isaac Newton. His statement of the laws of gravity, velocity and mass broke through the psychological barriers of his time to conceive of action at a distance, of force acting through empty space, so inconceivable to his predecessors. Newton's theory was worked out with austere mathematical purity, and it is typical of the quality of his mind that, in this great age of scientific images, he had no resort to visual diagrams or conceptual models. Newton's dynamic theory of cosmic structure was to satisfy the scientific mind for two centuries and more, and incidentally it marked the final separation of astronomy from astrology. By the later seventeenth century the universe was no longer seen as a closed system of spheres where astral forces radiated upon the Earth. In removing the Earth from the centre of the cosmos, science had shattered the classical model upon which astrology had been built, and Newton's dynamics rationalised the entire cosmic structure. Newton's work was the final vindication of empirical, mathematical astronomy, against the scholastic authorities of the past. Significantly the youthful Newton inscribed in one of his notebooks 'Amicus Plato, amicus Aristoteles, magis amica veritas' – 'Plato is my friend, Aristotle is my friend, but my greatest friend is truth'. Among the most notable casualties of the scientific quest for truth was astrology, which could not survive the new disciplines of observation, analysis, deduction and proof.

There was no conceivable physical evidence for the mechanism or forces through which astrology was supposed to operate, and by the end of the seventeenth century it had withered away. Astrology lost for ever it place in western intellectual life, although of course it survived in the popular imagination and has never disappeared.

The Earliest Printed Star Charts

It was in this period of revolutionary astronomical change that the star chart established itself as a scientific reference document comparable to the world map. The Copernican revolution, and the problems it threw up, re-focused attention on every aspect of the heavens, and demanded new data and new methods of observation, study and interpretation. This was the empirical revolution in astronomy, and its effect was to produce an outpouring of research and popular publications which described and portrayed the skies. The emergence and proliferation of published star maps after 1600 would have been unthinkable without the impetus provided by the developing revolution in astronomy. The paths of the planets were not, of course, directly shown on published star charts, for the chart would then be valid only for a period of months. Planetary positions were published in tables known as ephemerides, which the observer, professional or amateur, astronomer or astrologer, could then relate to his star chart. By the early seventeenth century there was a conscious sense that this was a new age of astronomy. Yet the earlier problem of designing a conceptual model of the starry sphere, a stellar reference system, was solved long before the problem of cosmic structure. The radical remodelling of the world map which accompanied the revival of classical geography had a crucial influence on the emergence of the star chart, for the appearance in Europe in the mid-1400s of a fully fledged map of the heavens was exactly contemporary with the earliest classically modelled, co-ordinated world maps. Individual constellation figures, some more, some less, scientific in style, had remained familiar throughout the Middle Ages, yet the most authoritative astronomical text, Ptolemy's *Almagest*, was not traditionally illustrated (except with geometric figures) and before the early sixteenth century no great interest had been shown in diagrams which located the stars and constellations on the vault of the heavens.

The earliest known genuine star map is essentially a new creation dating from *c.* 1440, the unique Vienna manuscript. It appears in an anonymous astronomical work *De Composicione Sphere Solide*, and it forms the prototype of an image that was to be replicated and elaborated hundreds of times over the next three centuries. The forty-eight Ptolemaic constellations are shown, with the stars numbered according to the catalogue in the *Almagest*. The maps are projected from the ecliptic poles, so that the ecliptic itself, adorned with the Zodiac figures, forms the maps' edges. The Milky Way is shown, and the 360 degrees of the ecliptic circle are individually marked, subdivided into blocks of 30 degrees, one block for each Zodiac sign. There is no bar of altitude or celestial latitude. The source, origin and purpose of this, the

Above
The Vienna manuscript, *c.* 1440. The earliest surviving genuine star map of the northern heavens, plotted within co-ordinates and projected from the ecliptic pole. The stars are marked with the numbers in Ptolemy's star catalogue. The origin of this map is unknown, but it became the pattern for all future star charts. Manuscript maps like this were probably circulating among Renaissance scientists for many years before a printed version appeared.

oldest two-dimensional star map, are all unknown. The scholar who drew
it (or if it is a copy, the scholar who drew the original) had made the crucial
transition from globe to map, and the lack of an altitude scale is suggestive:
while Islamic globes always bore the ecliptic or equatorial degrees on their
surface, the latitude degrees were almost always marked on a detached
meridian. A certain Islamic influence is detectable in the iconography of some
constellations, for example the scimitar in Hercules's hand, and the turban-
style crown or helmet worn by Cepheus. It seems virtually certain that the
unknown author of this planisphere did what had been inherently possible
for many centuries, but which no one had felt impelled to do: working from
a celestial globe, and perhaps from an astrolabe too, both probably Islamic,
he spread out the starry sphere to form the first paper star chart, and created
a conceptual model of the heavens which no human would never see, but
which would become permanent in western science. It may truly be described
as a map and not a picture, because its co-ordinate framework gives it a
measured structure in which each part of the map is precisely related to the
others and to reality.

The context and purpose of the Dürer planispheres are not known
with any certainty, but it is clear that they were the printed culmination of
half a century of thought and experiment among a group of intellectuals
centred on Vienna and Nuremberg. The unknown author of the Vienna

manuscript must have been involved in this group at an early stage, and
Martin Bylica of Cracow University was another associate. He commissioned
from the instrument-maker Hans Dorn the first European star globe, cast
in metal, very much on the Islamic model and strikingly similar to the
Vienna manuscript. The key figure was Regiomontanus, whose influential
epitome of the *Almagest* was completed in 1463 but not published until
1496. After studying in Vienna he was lecturing on Islamic astronomy in
Padua during the 1460s. He then travelled to Hungary and worked with
Martin Bylica, before settling in Nuremberg in 1471, where he taught and
published. It is known that he and Bylica were very familiar with Islamic
globes and astrolabes, and it seems certain that manuscript star charts were
circulating among these scholars. The moment had plainly arrived when
spherical geometry, Ptolemaic co-ordinates and the visual example of Islamic
instruments should all combine to produce the two-dimensional planispheric
star map. The first printed examples were designed as woodcuts by Albrecht
Dürer and printed in Nuremberg in 1515, and were immensely influential in
shaping the genre of the printed star chart over the following three centuries.

The cartouche at the foot of the southern chart explains that two
Nuremberg mathematicians, Johann Stabius and Conrad Heinfogel, plotted
the star positions on the charts, while Dürer was responsible for the artistic
design. An important point relating to this and all subsequent star maps is

Opposite & Above
Apian's star chart from *Astronomicon Caesareum*, 1540. Apian's map is obviously derived from Dürer's planispheres, but its novelty lies in its combination of the northern and southern spheres into one chart: the projection from the pole is extended south beyond the ecliptic, so that the south pole would eventually become a line encircling the whole map. In this way all forty-eight classical constellations can be shown. The lateral distortion becomes extreme, as can be seen from the length of the ship *Argo*. Nevertheless, it is surprising that this ingenious form of star map was not more widely imitated.

the distinction between the map itself and the star catalogue on which it was based. The materials of the Dürer star charts are thoroughly Ptolemaic: the constellations are the forty-eight listed in the *Almagest*, and the stars are identified by the numbers which Ptolemy gave them. The altitude of stars north or south of the ecliptic is unchanging, but the phenomenon of precession means that all celestial longitudes will shift westwards by one degree every seventy-two years. If Stabius and Heinfogel had used Ptolemy's co-ordinates from AD 150 all the Zodiac constellations would have been 19 degrees out of position. The Alfonsine tables of AD 1252 had updated Ptolemy, and the Nuremberg group further corrected those figures to allow for precession since that date. It would be open to any artist or publisher simply to copy the form of a celestial map from any available model, but if the aim was to produce a scientific document for the discerning reader, then up-to-date longitudinal positioning of the stars was essential. Dürer and his scientific editors were plainly aware of the classical tradition that the constellation figures should be portrayed as if seen from the outside of the starry sphere: the images are seen in rear-view and progress anti-clockwise around the heavens. Dürer has to some extent westernised his images; for example, Hercules now bears a club and a lion skin. In the striking marginal portraits Dürer acknowledges the sources and authorities of all astronomical works: Ptolemy, Aratus, Al-Sufi (his name latinised to Azophi) and Marcus Manilius. The last is the least familiar: a minor Roman poet who wrote a verse treatise on astrology, and whose work Dürer happened to know because the first printed edition of it had been issued in Nuremberg some years earlier.

The Dürer planispheres represent both a beginning and an end, or rather a culmination. They inaugurated a genre of publishing in which a scientific document sought to give delight through its artistry, a genre which was to flourish for 300 years and more. But it also forms the culminating link in an intellectual chain whose length is deeply impressive. Constellation figures, traceable in some cases to Mesopotamia in the second millenium BC, were placed within a geometric framework by Greek science, a framework perpetuated by Islamic scholars during centuries of medieval western neglect, and this tradition is then received by a north European artist and given an elegant graphic form which satisfies the quickened visual imagination of the Renaissance. The Dürer figures were copied by Peter Apian but cleverly combined into a single sphere for the star map in his monumental work *Astronomicon Caesareum*, 1540. This famous work was one of the most sumptuous productions of the first century of printing. Apian created a series of revolving paper models called volvelles, in which the movement of discs pinned at their centres simulated those of the celestial bodies. They were paper equivalents of the late medieval instruments, for instance simple equatoria or the much more intricate albion of Richard of Wallingford, which all aimed at finding planetary positions; the fact of their being printed on paper has often led to their being described as maps. Apian's work was more astrological than astronomical in purpose, and the book definitely inhabits the pre-Copernican world. The volvelles were used for casting horoscopes and they made ingenious toys for Apian's wealthy patrons.

Above
The constellation Scorpio, from
Alessandro Piccolomini's *De le Stelle
Fisse*, 1540. Piccolomini's book was
the first printed celestial atlas and it
was highly detailed. But its austere
style, lacking co-ordinates and the
constellation figures, limited its
popularity.

A New Era of Celestial Mapping Begins

Another class of astronomical image which was emphatically not new but
which was now transformed by the medium of printing was the portrayal
of the constellations. The first printed book of star maps was Alessandro
Piccolomini's *De la Stelle Fisse* of 1540, but the use here of the term map is
debatable: Piccolomini's stark star-groups, with no constellation-figures, have
an austere, scientific appearance; but without printed co-ordinates (although
they were clearly planned with co-ordinates to hand) they can no more be
located on the celestial sphere than Al-Sufi's could be four centuries earlier.
A definite advance appears in Giovanni Gallucci's *Theatrum Mundi* of 1588
where the constellations are drawn within a framework of co-ordinates and a
geometric projection. Building on the foundations laid by Dürer and Gallucci,
the printed star charts of the next three centuries bear eloquent witness to
the evolution of both the astronomy and the publishing taste of their day.
New constellations were added, the Moon was mapped, and the Copernican
planetary system gained universal acceptance. Ptolemy's star catalogue was
at last superseded and a series of landmark celestial atlases set ever new
standards of fullness, accuracy and gracefulness. Geographical exploration,
politics, religion, and the artistic styles of the time all left their mark on the
evolving star map. Comets, eclipses, solar transits and nebulae were recorded
and publicised as the star chart entered the mainstream of scientific life and
of popular publishing. Like terrestrial maps and like other forms of technical
publication (architecture for example, or medicine), star charts developed
in a dual context. On the one hand, leading astronomers such as Hevelius,
Flamsteed and Bode were advancing the science of astronomy through
innovative charts based on original research, while on the other hand
resourceful publishers such as Cellarius and Ottens were marketing well-
designed but derivative star maps for a wider, educated audience. Elegant,
compact and intellectually satisfying, maps of the northern and southern

heavens became almost as familiar as the map of the world, and essential features in any library.

Contemporary with the appearance of the first printed star charts were the earliest globes, both celestial and terrestrial. Knowledge of the sphericity of the Earth was widespread in the Middle Ages (contrary to modern popular belief) but the impulse to make three-dimensional models of the world from the late fifteenth century onwards stemmed directly from the great period of exploration and the sense of the world encompassed. The classical concept of the celestial sphere as enclosing the terrestrial was revived, and complementary models of the Earth and the heavens were produced. Islamic scholars and craftsmen had never made terrestrial globes, but had for centuries made celestial ones, which by the sixteenth century were well known and admired in Europe. These had been skilfully made by the technically difficult process of metal casting. When German craftsmen

Opposite, below
Gallucci's *Theatrum Mundi*, 1588. An early woodcut star atlas illustrating each of the constellation figures, with a catalogue of their stars. Gallucci followed the star catalogue of Copernicus rather than Ptolemy. In order to avoid the changes caused by precession, Copernicus measured celestial longitude not from the vernal equinox, which is itself a slowly moving point, but from the first star in Aries. Thus the movement of the entire celestial sphere leaves all co-ordinates constant in relation to each other. Gallucci has used a trapezoid frame to indicate that these maps are sections of the celestial sphere, narrowing towards the poles.

Right
Johann Schöner's celestial globe, *c.* 1533. The first printed celestial globe. Schöner was the most influential early globe-maker, establishing Nuremberg as the European centre of the craft, and setting the pattern for pairing celestial and terrestrial globes. This globe appears to be the model used by Hans Holbein in his painting *The Ambassadors*.

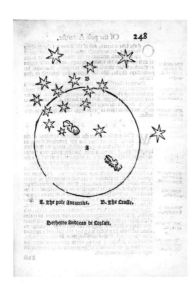

Above
South polar stars from Richard Eden's *Decades of the New World*, 1555. The Southern Cross and the Magellanic Clouds, drawn from an account of Magellan's voyage to latitude 55 degrees south during his circumnavigation of 1519–1522.

Above
Navigation, from Pedro da Medina's *Regimiento de Navigation*, 1563. A mariner uses a cross-staff to determine the altitude of the pole star above the horizon, and hence his latitude.

devised the simpler method of moulding papier-maché around a carved ball of metal or stone, the moment was ripe for the production of globe pairs, matching the Earth to its cosmic setting among the stars. The globe would always be more costly to make than a map, and for that reason it acquired a symbolic value as an emblem of scholarship married with wealth; as such, we encounter it in pictures such as Holbein's *The Ambassadors* of 1533. The novelty of the celestial globe even in Renaissance Italy may be gauged from Raphael's great fresco *The School of Athens* c.1510, where an unidentified philosopher can plainly be seen holding a celestial globe, but where the artist has made no attempt to paint the actual stars or constellations realistically; unlike Holbein two decades later, he had no model to work from, or perhaps felt that the figure of the pictorial globe would still be unfamiliar. The ultimate use of maps of the heavens (and of the Earth) as displays of power may be seen in the vogue for mural decorations such as those in the Villa Farnese at Caprarola, where the possession of a private image of heaven seems to symbolise not merely the Farnese family's grasp on the things of this world, but their equality with the mythical heroes of the skies.

At the other extreme from the large globe as intellectual library furniture were the miniature models designed for the gentleman's, and perhaps for the mariner's, pocket. A terrestrial globe just 2 or 3 inches in diameter nestled inside a spherical case, on whose inner concave surface were drawn the northern and southern heavens. Largely a novelty testifying to the eighteenth-century vogue for the science of astronomy, they might still have been some practical use for demonstration purposes on sea or land. From ancient times the stars had, of course, been fundamental to marine navigation, the Plough (Ursa Major) and the pole star forming permanent markers for all seafarers in the northern hemisphere. Although it was supplemented in the fifteenth century by compass and chart, stellar navigation was given even greater importance by seaborne expansion beyond European waters. The relationship between the celestial pole and the observer's horizon in fixing one's position was easily understood: the elevation of the pole at any point of observation is equal to that point's angular distance from the equator – its latitude. This theory had been expressed three centuries before in Sacrobosco's *De Sphaere Mundi*. Yet in order to put into effect this fundamental rule, an established latitude system was clearly essential, both celestial and terrestrial. Celestial navigation in this precise sense does not, therefore, pre-date the late fifteenth century when co-ordinate systems became familiar and began to appear on sea-charts, and when instruments for determining altitudes began to proliferate – the quadrant, the cross-staff, and the mariner's astrolabe. The last was much simpler than the astronomer's astrolabe, and was used to measure the altitude of the Sun as it crossed the noon meridian, a technique increasingly employed as mariners voyaged southwards out of sight of the pole star.

It is sometimes thought that the age of exploration and consequent demands on navigation stimulated astronomers to devise new instruments, new observational skills and new theories, but there appears to be little direct correlation between the two. Navigational advances proceeded very much

from pragmatic necessity, until the late seventeenth century at least, and sprang from within the maritime community. The revolution in empirical astronomy associated with Tycho, Kepler and Galileo was purely intellectual and scientific, having no connection with maritime enterprise, although of course their work would later have profound repercussions for navigation and geography generally. It was to be many years, the late seventeenth century at the earliest, before a new professionalism in maritime training would require the seafarer to have a thorough grasp of astronomy before he could be called a navigator.

Changes in instrumentation during the seventeenth century led to a shift from the use of the ecliptic co-ordinate system to the equatorial system, so that stellar latitude and longitude were directly in line with those on Earth. There is no great intellectual significance in the change: it was simply found more convenient. Nor did the change occur suddenly, for throughout the period 1550–1750 either system might be used, so that one has to look carefully at any given chart or globe to determine whether it is projected from a celestial ecliptic or equatorial pole. It means that a star map entitled

Right

Remmet Backer's star chart, an eighteenth-century printing. An ingenious application of the Mercator-type projection to the celestial map. The heavenly sphere is conceived to be peeled back from the poles to form a cylinder, with the poles becoming complete circular bands. The cylinder is then unrolled to yield a rectangular map 180 degrees × 360 degrees, exactly analogous to the Mercator world map. The centre of the projection is the vernal equinox point, where the celestial equator crosses the ecliptic (again analogous to the centre of the world map where the Greenwich meridian crosses the equator).

As a reference model displaying the relative positions of the stars and constellations, it is eminently useful; but like the Mercator world map, it is remote from the shape of the sphere, and from anything the human eye can see.

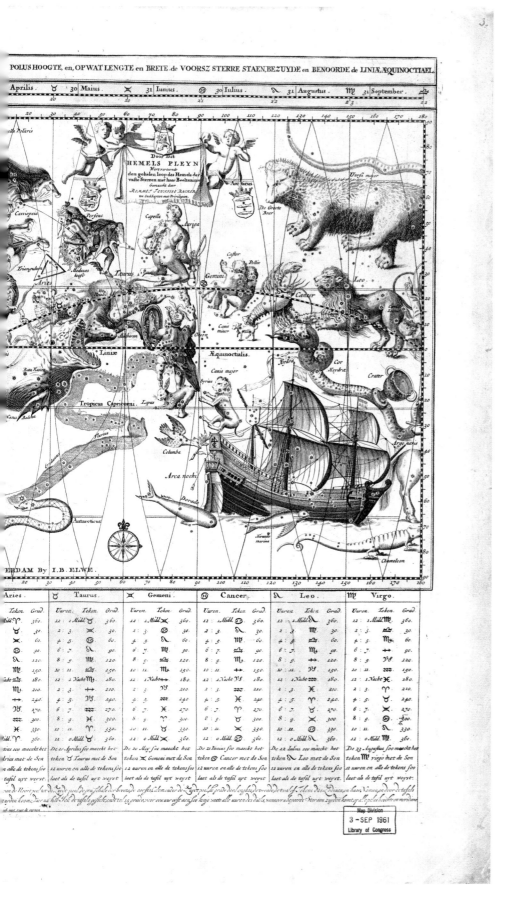

'Northern Celestial Hemisphere' may not show the northern skies in our sense because the ecliptic will form the map's edge and extend to a point 23 degrees south of the equator. It was not until the late eighteenth century that the equatorial system became virtually universal. In both systems celestial longitude was measured from the vernal equinox point, latitude with reference to the relevant baseline – either the equatorial or the ecliptic.

The other technical matter which the makers of star charts had to consider was projection. In the case of world map projections, the sixteenth century saw intense and varied experimentation as cartographers sought to portray on paper their radically expanding world. One of the controlling factors was the natural desire to place Europe at the centre of the map and then to reach out from the old world to show the new discoveries to the east and the west. But the mapmaker always felt it essential to show the entire world, so that his problem became complex and capable of a great variety of solutions. In the case of the star chart there was far less experimentation, mainly because celestial chartmakers attempted to portray only one half of the sky; so two maps would be paired, one of the northern sky and one of the southern. The central feature was always conceived to be the celestial pole around which the heavens revolved. It was natural to choose the pole as the centre of the projection and to chart the stars as if the observer were looking down on the celestial sphere from a point high above the pole. The process of drawing the chart may then be visualised as if viewing a dome on which the stars appear, and which must then be spread out and transferred to a flat surface.

It will be easily seen that as one progresses away from the central point of tangency, space becomes distorted and a decision must be taken how far from the pole to continue the chart. The solution is to stretch the map progressively away from the pole so that an entire hemisphere down to the equator or even the tropic of Capricorn may be shown. The problems of projecting the celestial sphere on to paper are identical to those of projecting the spherical Earth, but at an early stage the polar projection established itself as the norm for star charts and there were few departures from this rule. Perhaps the most interesting departure is the map originally designed in 1684 by Remmet Backer (see page 120) which is an ingenious equivalent of the Mercator projection world map. Maps of the entire northern or southern heavens were not observational instruments; they were convenient conceptual models, or reference charts, and were accepted as such by their scientific users. But in general celestial mapmakers felt that no purpose would have been served by experimenting with a variety of new celestial projections, since the pragmatic fact that the starry sphere was never seen in its entirety was accepted as according with our senses and our reason. The polar projection is used to expand the two-dimensional space available to the draughtsman. The maps of individual constellations which filled celestial atlases were effectively regional star charts, analogous to maps of individual countries in a world atlas, in which the problems of projection were correspondingly less severe.

Historical Changes in Star Mapping

Celestial charts were undoubtedly more conservative documents than most other forms of map, so that the broad structure of the first manuscript planisphere of 1440 is still dominant in an eighteenth-century star chart. But within this framework of strong continuity, there was a process of variation, elaboration, and refinement, as well as short-lived experiments and a few permanent transformations. Since the appearance of Ptolemy's star catalogue around the year AD 150 no astronomers for 1,400 years had ventured to add to his forty-eight constellations. Challenges to this status quo obviously arose first with the voyages of discovery, and second with the movement to re-catalogue the stars which began with Tycho. From around 1450, Portuguese mariners in search of a sea-route to the east voyaged steadily further south along the coast of Africa until Bartolemeu Dias rounded the Cape in 1489. Obviously these sailors lost sight of almost all their familiar stars, and they must have quickly identified others to sail by. Yet from all the Portuguese and Spanish seafaring enterprise of the sixteenth century, almost no records of star sightings or new constellations have survived. Amerigo Vespucci's narratives of his Atlantic voyages, published from 1504 onwards, contained small figures of a handful of new stars, while the remnants of Magellan's crew told of two detached clouds of light visible far to the south and apparently detached from the Milky Way – the Magellanic Clouds. The Portuguese navigator Alvise Cadomosto reported near the mouth of the river Gambia, at 13 degrees north, that the Pole star sank so low that it touched the sea, while to the south an outstanding group of stars in the form of a cross appeared – the first reference to the Southern Cross. These are all the direct astronomical results of these pioneering voyages. We must assume that the Portuguese who were sailing the Cape route and the Indian Ocean throughout the sixteenth century did in fact identify new guiding-stars, but that some motive – perhaps a desire for secrecy – prevented their publication. It was not until 1595–96 during Frederik de Houtman's voyage via the Cape to Java, that his Dutch compatriot, the navigator Pieter Dircksz Keyser, systematically formed a host of southern stars into twelve new constellations. He chose to name most of them after exotic beasts, some mythical like the Chameleon and the Phoenix, others native to the newly explored territories of South Asia, like the Toucan, the Bird of Paradise and the Flying Fish. On Keyser's return to Europe these were published in textual form with the star co-ordinates, and in graphic form on celestial globes made by Hondius and Blaeu in the last years of the sixteenth century. They became more generally known through Johann Bayer's great star atlas *Uranometria* of 1603, and they have become permanent.

This mood of innovation seems to have communicated itself to other astronomers, for in 1624 Jakob Bartsch, who was Kepler's son-in-law, formed a number of previously detached stars into new northern constellations, the Giraffe, the Dove and the Unicorn. In 1687 Johannes Hevelius of Gdansk designated a further seven groups, almost all northern, including the Lynx, the Fox and the Hunting Dogs. The last permanent block of constellations to be designated by an individual astronomer was the creation of Nicolas

Bayer's *Uranometria*, 1603. Bayer's was the first engraved star atlas, and his maps have noticeably more precision than the earlier woodcuts of Gallucci. Bayer used the best contemporary star catalogue, that of Tycho, and he included the newly designated southern constellations of Pieter Dircksz Keyser. Bayer made the rational decision to portray the constellations as seen from Earth. Another innovation introduced by Bayer was to classify the stars within each group according to their magnitude by using the letters of the Greek alphabet, so that for example Mirfak, the bright star in Perseus's side, becomes alpha Perseus. This concept soon became indispensable to astronomers as a stellar reference system.

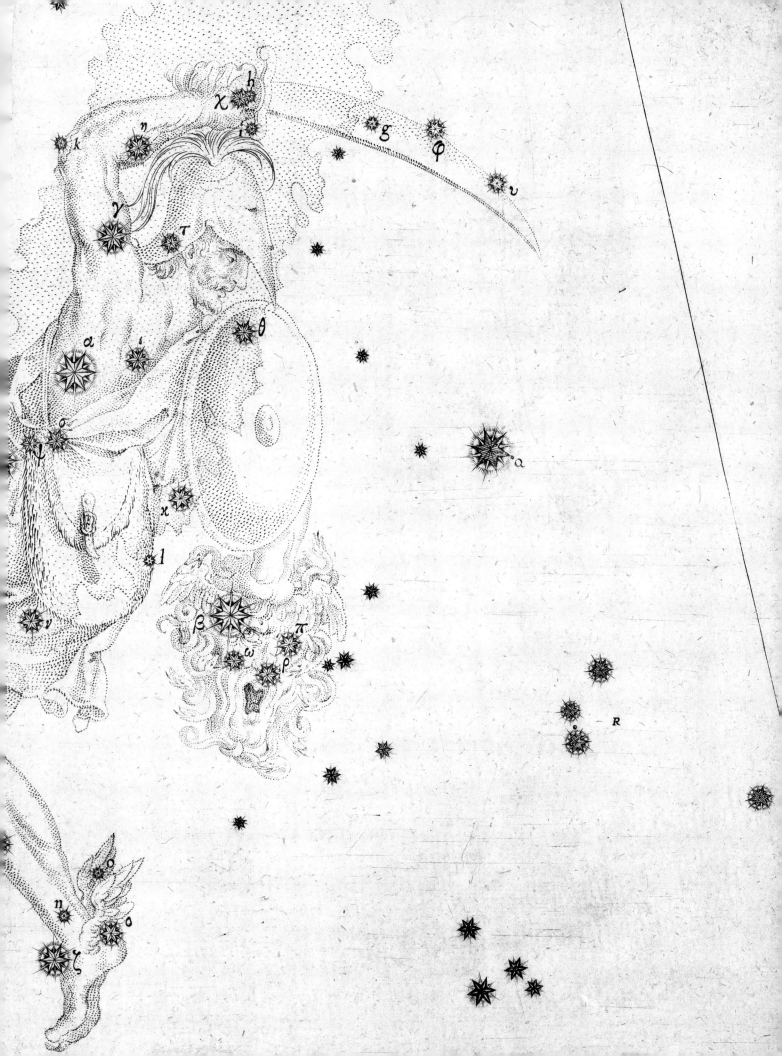

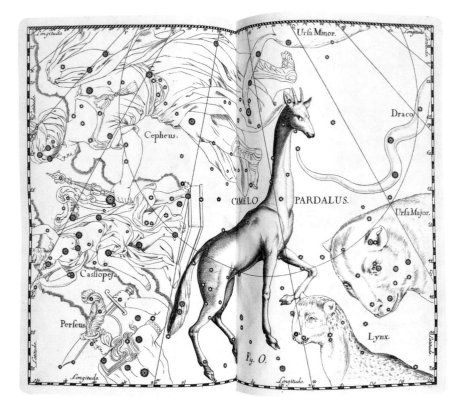

de Lacaille who spent the years 1751–52 surveying the southern sky from the Cape of Good Hope. Working consciously in the spirit of eighteenth-century science, Lacaille turned his back on traditional animal motifs and chose instead technical subjects: the Clock, the Compass, the Telescope, and so on.

Although Lacaille was the last individual to leave the products of his imagination permanently in the map of the skies, there were many other attempts to create new constellation, some of them fascinating, but which failed to win recognition from the world of astronomy; these attempts reflect the intellectual and sometimes the social context in which astronomers were working. One prominent theme or motive in the design of new constellations was the quest for political favour or patronage, understandable perhaps in an age when kings and nobles were the only possible source for the endowment and support of observatories. The French astronomers Pardies and Royer proposed a Fleur de Lys and a Sceptre to honour Louis xiv. Godfried Kirch invented Swords, Sceptres and Orbs in honour of various German royal houses. In Restoration England, Edward Sherburne and Edmond Halley devised respectively a Royal Heart and a Royal Oak in commemoration of Charles I and Charles II. Given their strongly partisan character, none of these symbols was likely to win permanent acceptance among the international scientific community. Perhaps the most ambitious of all the products of this political astronomy was the plan of Erhard Weigel in Jena in the 1680s to replace all the classical constellations with the coats of arms of the ruling houses of Europe.

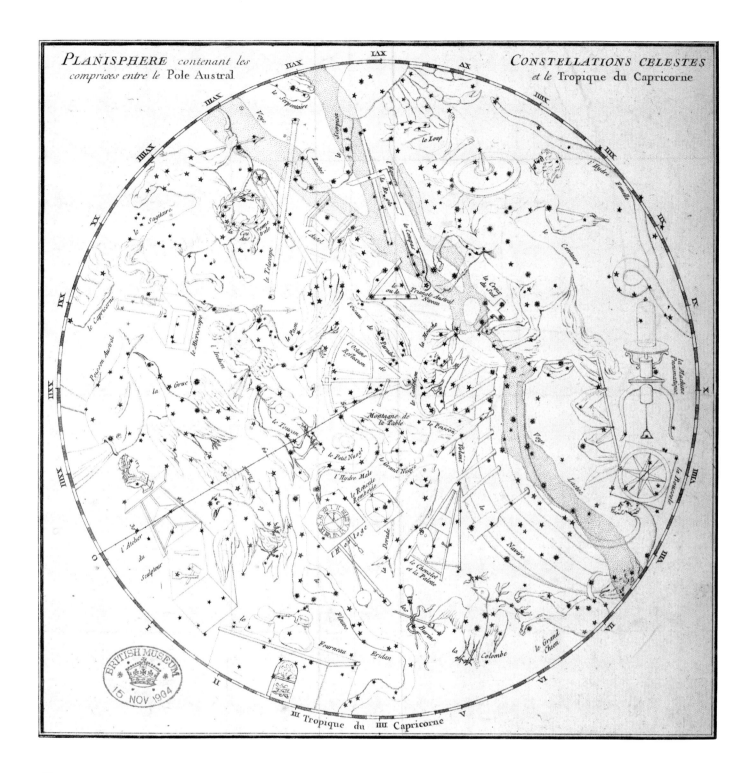

Above

Lacaille's constellations, 1752.
The French astronomer Nicolas de
Lacaille spent the years 1751–52
surveying the southern skies from
the Cape Observatory. He designated
thirteen new constellations based
on scientific instruments. His were
the last new constellations to win
universal acceptance.

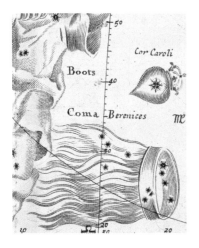

It is easy to forget that there was a serious motive behind the creation of
new constellations: the desire for completeness and the desire for new and
more accurate star catalogues. The classic catalogues of Ptolemy and Tycho
made reference only to some 1,000 stars grouped into the familiar forty-eight
constellations; but what of the thousands of other stars which lay between
constellations, the so-called unformed stars? How were they to be catalogued
with no star-group or sky-region to locate them? There were substantial
empty spaces, even in the northern sky, so that for example Hevelius's motive
in creating the constellation Canes Venatici – the Hunting Dogs – was to
impose some plan on the area of sky between Ursa Major and Boötes. This
may be thought of as analogous to the increasingly large scale of terrestrial
maps, so that more detail could be represented.

No less radical than the political constellations, but also doomed to
failure, was the imaginative scheme to de-paganise the heavens, to replace
the mythological and animal figures with Christian and Biblical subjects.
This project took two forms; on the simpler level the 1655 celestial maps of
Wilhelm Schickard reproduced the familiar figures, but added brief texts
suggesting ingenious Biblical equivalents for them. Cepheus, for example,
is compared to Solomon, Perseus to David, Draco to the dragon in Revelation.
This attractive idea was fully worked out and given dramatic visual expression
in the *Coelum Stellatum Christianum* of Julius Schiller published in 1627, who
devised a radically Christianised map of the heavens. Schiller's southern
hemisphere was transformed into a cavalcade of Old Testament subjects – Job
takes the place of the Indian and the Peacock, the Centaur is transformed

Right
Weigel's political constellations,
1688. Many new constellations
were devised by astronomers to
honour royal patrons, but none of
these innovations was as radical
as Erhard Weigel's scheme for
replacing the entire face of the
classical heavens with the coats
of arms of Europe's royal houses
– the German eagle, the French
fleur de lys, and so on. Weigel
was a professor at the University
of Jena, and a distinguished
mathematician, and we must
assume that his published proposal
was in earnest. However, like
most politically inspired proposals
for constellations, it failed to win
acceptance.

Next Spread
Cygnus as the Holy Cross, 1627.
Perhaps the most intriguing
attempt to replace the traditional
constellations was that of the
German scholar Julius Schiller.
Allotting the northern heavens
to the New Testament and the
southern to the Old Testament,
Schiller found biblical and
ecclesiastical replacements for
all the classical star-groups. Their
failure to win acceptance even in
Catholic countries demonstrates
the tenacity of classical traditions
in European art and science.

into Abraham and Isaac, and so on. The northern heavens are filled with
New Testament and Christian imagery: Cassiopeia becomes Mary Magdalen,
Perseus St Paul, while the twelve Zodiac signs are conveniently replaced by
the twelve apostles. Schiller's maps reached a high standard of detail and
precision; they were certainly not intended to be capricious or ephemeral,
yet his innovations failed to persuade his contemporaries (perhaps because
of the heavily Catholic nature of his imagery), and his work is now only an
intriguing footnote in the history of astronomy. It is difficult not to sympathise
with Schiller, for the desire to Christianise the heavens was neither an eccentric
nor a controversial ambition, and he might legitimately have expected to
succeed in Christian Europe. Schiller's constellations were contemporary with
the elaborate scenes of heaven painted on the ceilings of baroque churches
in Catholic Europe, and his maps underline the paradox of the survival
throughout the Christian centuries of the pagan, animistic sky-figures.

The Classical Age of the Star Atlas

Of far greater permanent value than these experiments was the series of
great celestial atlases, each of which in turn set new standards of fullness and
accuracy. The woodcut atlases of Piccolomini and Gallucci were extremely
important in their way, but a glance at Johann Bayer's atlas *Uranometria*,
published in Augsburg in 1603, reveals immediately a different order of
achievement. The medium of copper-plate engraving created a much more

Right

The Christianised heavens of Schiller, 1660. Schiller himself did not publish general maps of his Christianised heavens, but some years later Andreas Cellarius did. Here, the southern hemisphere is filled with Old Testament figures such as Job, Aaron and the angel Raphael, while the north shows St Paul, Jerome and others. The Zodiac figures along the ecliptic have been transformed into the twelve apostles. The map and its twin adopt an unusual structure: instead of being projected from the poles, they centre on the equinoxes. The ecliptic bisects the map instead of encircling it, and each hemisphere shows part of the northern and the southern sky. These maps are exactly analogous to the twin-hemisphere world map that was prevalent during the seventeenth century.

precise image than the woodcut, and, in the hands of the skilled engraver, the modelling of the constellation figures has a rich, sculptural quality. Bayer had access to the new star catalogue of Tycho, which could claim to be the first new star survey of the modern age. Although Bayer does not show a dramatically increased number of stars compared with Gallucci, each one is precisely located and graded for magnitude, and it was in this atlas that Bayer introduced for the first time the system of ordering the stars within each constellation by the letters of the Greek alphabet. This excellent principle was not universally imitated until the following century.

Johannes Hevelius's atlas *Uranographia* of 1687 consolidated the approach of Bayer, and first showed the seven new constellations of his own devising. Even at this late date, when the telescope had been in existence for some seventy-five years, Hevelius preferred non-telescopic instruments to sight the stars. During a visit from Edmond Halley to his observatory in Gdansk, Hevelius attempted to demonstrate that he could take more accurate stellar positions than Halley could with a telescope. It was Hevelius too who had the distinction of publishing the first atlas of the Moon, *Selenographia*, 1647. Constructed from his own observations, this time of course with a telescope, it displayed for the first time the complexity of Moon's topography, although it perpetuated certain myths such as the existence of lunar seas. Few of the place-names proposed by Hevelius became permanent; indeed, one of the most striking aspects of his maps is the elaborate analogy he built up between the topography of the Moon and that of the Earth, with the Mediterranean,

Above
Hevelius: the Moon from
Selenographia, 1647.

Right
Argo from Hevelius, *Uranographia*,
1690. Johannes Hevelius's star
catalogue of more than 1,500 stars
was the most comprehensive of its
time, and it formed the basis of his
great celestial atlas. He designated
seven new and permanent
constellations, including the Sextant
(the first scientific subject) and
the Shield of Jan Sobieski, King of
Poland. He included others which
had a short life-span, such as
'Charles' Oak', seen here, which he
copied from English astronomers.
Unlike Bayer, Hevelius reverted to
the classical reverse orientation of
the constellations.

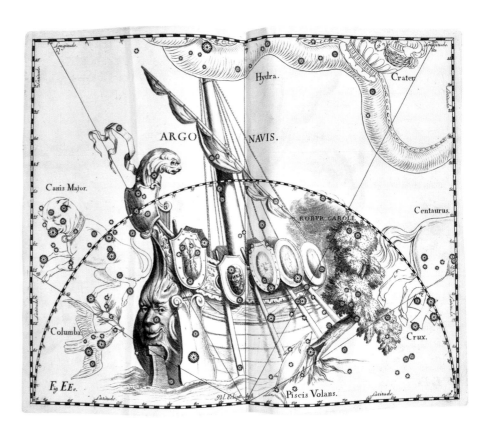

North Africa and Asia Minor dominating the Moon's visible face. It is to the Jesuit astronomer, Giambattista Riccioli, an ardent opponent of Copernicanism, that we owe most of the familiar lunar names.

The foundations for a new area of research vital for the development of astronomy were laid between 1770 and 1785 when Charles Messier began the observation and charting of nebulae. What began as an offshoot from Messier's interest in comet-hunting – a widespread addiction among eighteenth-century astronomers – culminated in the identification of 109 star clusters, nebulae or galaxies, and the recognition of the importance of these bodies in understanding the heavens. Messier himself did not embody his researches in an atlas of nebulae but he had the rare distinction of identifying a whole new class of objects in the sky for future mapmakers to chart. Comet-hunting also gave rise to a positive torrent of publications showing the shape and position of the latest visitor from space. Some of these were ephemeral broadsheets, while the major atlas publishers vied with each other to design elaborate diagrams of comets passing the Earth against the starry background.

New standards were set by John Flamsteed, the first Astronomer Royal in Britain, whose *Atlas Coelestis* of 1729 contained maps of the twenty-five constellations visible from Greenwich. Based on his own catalogue of nearly 4,000 stars, it was unprecedented in its detail, and the graceful, classically modelled engravings were designed by one of the foremost artists of the day, Sir James Thornhill, who had created murals in many of Wren's great buildings

Right

Seutter: The comet of 1742.
For eighteenth-century scientists,
comet-hunting was a favourite sport.
Although comets were mysterious
and apparently random in their
appearance, Halley had demonstrated
that they were indeed part of the
solar system. Any new comet was
eagerly studied for the light it might
shed on planetary mechanics.
Map publishers capitalised on this
interest by issuing numerous charts
of the paths of comets. This map by
Mattheus Seutter, of the comet that
was visible from 13 March to 15
April 1742, is unusually imaginative
in design, showing the comet's path
across the entire celestial sphere, and
also in more detail its position during
its thirty-three days of visibility in the
constellations Draco and Cepheus.

Next Spread

Flamsteed's *Atlas Coelestis*, 1729.
Flamsteed was the moving spirit
behind the foundation of the Royal
Observatory at Greenwich. Ten
years after his death this atlas was
published from his catalogue of
3,000 stars visible from Greenwich. It
was by far the fullest star catalogue
published to that date, throughout
the eighteenth century. The elegant
engravings were executed by James
Thornhill, and Flamsteed followed
Bayer's Greek-letter system to
catalogue the bright stars.

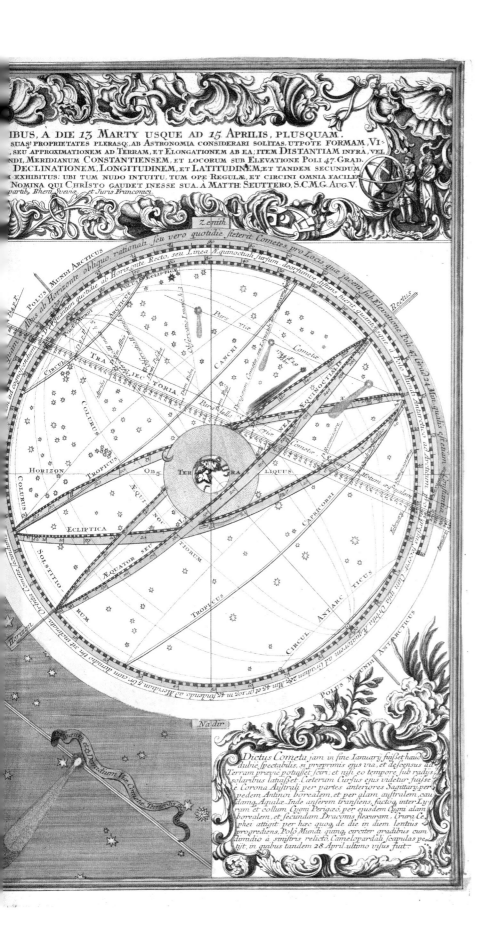

IBUS, A DIE *13.* MARTY USQUE AD *15.* APRILIS, PLUSQUAM.
SUAS² PROPRIETATES PLERASQ: AB ASTRONOMIA CONSIDERARI SOLITAS, UTPOTE FORMAM, VI-
SEU APPROXIMATIONEM AD TERRAM, ET ELONGATIONEM AB EA; ITEM DISTANTIAM INFRÀ, VEL
NDI, MERIDIANUM CONSTANTIENSEM, ET LOCORUM SUB ELEVATIONE POLI 47. GRAD.
DECLINATIONEM, LONGITUDINEM, ET LATITUDINEM, ET TANDEM SECUNDUM
EXHIBITUS: UBI TUM NUDO INTUITU, TUM OPE REGULÆ, ET CIRCINI OMNIA FACILES
NOMINA QUI CHRISTO GAUDET INESSE SUA. A MATTH: SEUTTERO, S.C.M.G.AUG.V.
partib. Rhena Suevia, et Juris Francomci.

*Dictus Cometa jam in fine Ianuarij fuisset haud
dubiè spectabilis, si præprimis ejus via, et descensus ad
Terram prævie potuisset sciri; et nisi eo tempore, sub radijs
solaribus latuisset. Cæterum Cursus ejus videtur fuisse
è Corona Australi per partes anteriores Sagittary, per
pedem Antinoi borealem, et per alam australem, cau-
dam, Aquilæ. Inde anserem transiens, facto inter Ly-
ram et collum Cygni Perigæo, per ejusdem Cygni alam
borealem, et secundam Draconis flexuram. Crura Cephei attigit per hæc quoq, de die in diem lentius
progrediens, Polo Mundi quinq, circiter gradibus cum
dimidio à sinistris relicto, Camelopardali scapulas pe-
tijt, in quibus tandem 28 April. ultimo visus fuit.*

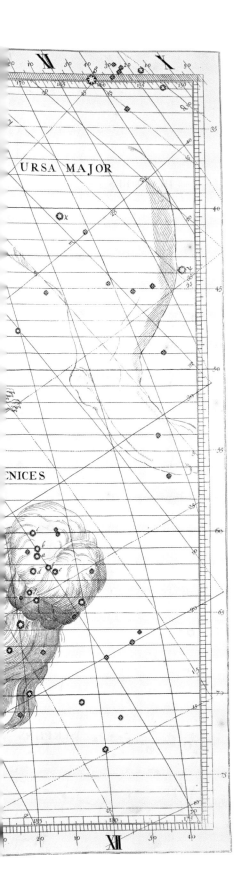

including St Paul's Cathedral and the Greenwich Hospital. The starting-point for Flamsteed's lifelong observations has often been retold: that in the quest for a means of determining longitude – a vital but elusive prize among the maritime nations – astronomers had suggested that the heavens might be used as a natural clock. The theory was that, for example, regular changes in the size and position of the Moon could be observed and tabulated at a standard meridian. Mariners at sea might then compare what they saw with these standard tables, and calculate the time difference between the two, each hour's difference representing 15 degrees of longitude. The theory was perfectly sound, but when Flamsteed was asked for his opinion, he replied that the accuracy with which the movements of the Moon had hitherto been measured was totally inadequate to the demands of this scheme.

This was the background to the establishment of the Royal Observatory, within which Flamsteed set out to create a new body of astronomical data for the new age of science. The striking thing about Flamsteed's star catalogue and his atlas was that they were derived from the first telescopic sky survey, since those of Tycho and Hevelius had both been optical only, using naked-eye sightings with large quadrants. Flamsteed's instruments were equipped with micrometers, and he achieved an accuracy of 10 seconds of arc, improving on Tycho's work by a factor of more than 10. The value of his observational work is clearly demonstrated by the eagerness with which the great theoretical scientists of his day, especially Newton, demanded access to it. Newton regarded the data which Flamsteed was accumulating as fundamental to his own continuing work on the problems of celestial mechanics, and moreover as public property since Flamsteed was a salaried public servant. In a bitter feud with Flamsteed, the forceful and irascible Newton virtually seized and published the data against Flamsteed's will. As with Tycho, it is interesting that the publication of celestial maps was not a priority with Flamsteed. He died in 1719 having completed virtually all the necessary observations for a new star catalogue, which appeared in three volumes in 1725. Finally his widow and one of his assistants published the atlas in 1729, as a memorial to the great observer and as an embodiment of his work which the non-specialist could appreciate.

The level of completeness sought by Flamsteed was extended to the rest of the heavens in the maps of Johann Bode. In his *Vorstellung der Gestirne* of 1782, Bode had built on Flamsteed's work, but extended it to cover the southern skies, and produced an atlas of around 5,000 stars. But it was Bode's later *Uranographia* of 1801 which represents the high-point of pictorial star atlases, showing more than 15,000 stars in a series of clear and vigorous engravings, which took star maps for the first time to the limit of naked-eye visibility and beyond. Even with the multitude of recently designated constellations, Bode's atlas is remarkable for the huge number of stars which lay between the figures, a fact which led to his innovative idea of constellation boundaries.

Within a few decades of the appearance of Bode's atlas, which had many imitators, astronomers began to feel that the pictorial representation of the constellations was a distracting anachronism, and during the nineteenth century a new, austere and functional form of star map appeared which

CAPUT HYDRÆ

Lactea

OFFICINA

TYPOGRAPHICA

Tropicus

PYXIS
NAUTICA

ET

LOCHIUM FUNIS

Markeb

variabilis

Previous Spread
Bode: *Uranographia*, 1801. Until
the later nineteenth century Bode's
remained the most comprehensive
celestial atlas ever published,
the first to attempt a complete
representation of all 15,000 naked-
eye stars. Particularly striking is the
number of unformed stars, that is
those lying outside any constellation.
It was in order to catalogue these
satisfactorily that Bode proposed
the novel step of designating
constellation boundaries, effectively
defining regions of the sky. This
system was to become fundamental
to astronomy, although the fixing of
the boundaries occupied many years'
experiment and adjustment.

dispensed with them. Pictorial star maps did not vanish entirely, but they became confined to popular scientific publications, while professional astronomers no longer used them. Agreed constellation boundaries were plotted which divided the sky into recognised regions, and international astronomical societies co-operated to call a halt to the arbitrary invention of new constellations. Thus the process of artistic elaboration of the sky map, which had begun in the printed medium three centuries earlier with Dürer, was ended in the scientific realm, though not in the popular mind.

The artistic form of the published constellation figures naturally varies with the style and taste of their time: they might be noticeably Islamic or Christian, medieval or Renaissance, classical or baroque. Provided that the stars themselves were shown in their correct mutual relationships, the actual style of the figures was naturally in the hands of the artist. One might expect an enormous variety of style and detail in the constellation images, yet through many centuries of pictorial star charts there remains a strong element of continuity, as though the artists were always aware that they were working within a long-established tradition. There was one stylistic problem which greatly exercised the makers of star charts over the years, namely whether the constellations should be pictured from the front or the rear. In his instructions on making a celestial globe, Ptolemy had argued logically that since we must see the model globe from the outside, and the Earth is conceived to lie at its centre, the constellations must appear as they would from beyond the starry sphere, that is from the rear. But should this apply also to a flat map, about which Ptolemy gave no instructions? If followed, it produces the obvious problem that the star groups do not appear as seen from the Earth, but as mirror-images, in which left and right, or front and back, are reversed. In the case of a lion, a snake or a bull, it scarcely matters if they are seen from one view or another, but in the human subjects, Perseus, Andromeda, Hercules and so on, these figures unquestionably appear odd and less attractive when they present their backs to us.

More important, it has the effect that it appears to throw the motion of the Zodiac into an anti-clockwise direction, instead of the clockwise progress observed from the Earth. The medieval illustrations to Aratus or Hyginus were completely outside the Ptolemaic tradition, and they worked in the most natural way, showing the figures from the front. Curiously the Islamic globe-makers and astronomical illustrators such as Al-Sufi, who were steeped in the Ptolemaic works, nevertheless chose to ignore him on this point and depict the constellations from the front. When the first star charts were drawn in the fifteenth century, in the context of the revival of classical science, their authors interpreted Ptolemy's principle as indeed applying to star maps, and this justified the rear-view which Dürer adopted in his seminal work, copied no doubt from the Vienna manuscript. During the next 300 years, this was the model which prevailed, although some mapmakers, Flamsteed for example, took the more pragmatic view that the constellations should be depicted as seen from the Earth, and that the classical rear-view was pedantic and lacking in sense. It is noticeable that both the newly designated

Right

Ottens: The southern sky, 1729. Another chart designed by the German mathematician Johann Doppelmayr, this shows clearly the southern group of exotic constellations (the Indian, the Peacock, the Toucan, etc.), and some that did not become permanent, such as the Royal Oak and the Fly. The views of the European observatories, including Greenwich, indicate that the map was aimed at pleasing an educated lay public, rather than for use by professional astronomers.

and the experimental constellations, those of Keyser and Schiller, took the opportunity to break with the classical convention.

The existence of this problem and the prevalence of the classical norm illustrate the conservatism that marks the long history of the star chart. Just as medieval artists, Islamic or Christian, had taken a non-scientific delight in the constellation figures, so even in the scientific climate of the seventeenth and eighteenth centuries there were publishers who were motivated more by the aesthetics of the market-place than by a desire to serve science. The most elaborate and famous celestial atlas of the seventeenth century was issued by an author unknown to the history of astronomy. Andreas Cellarius's *Atlas Coelestis* (Amsterdam 1660) contained dozens of imaginatively designed plates showing the classical and Christianised heavens, the Ptolemaic and Copernican planetary systems, and experimental projections designed to simulate a three-dimensional view of the Earth and the heavens. Cellarius's concept of a richly illustrated collection of maps combined with astronomical diagrams, aimed at the wider market outside the scientific professions, was imitated by several eighteenth-century map publishers, notably Homann in Nuremberg and Reiner Ottens in Amsterdam.

But undoubtedly the most visually powerful and frankly artistic star charts ever made were those of the Venetian globe-maker Vincenzo Coronelli (1650–1718). Published in the form of 'gores', which are the oval sections made to be cut out and pasted to a globe, Coronelli's constellation figures are triumphs of late baroque art: the sculptured forms seem replete with gigantic strength, and the darkly etched engravings create a series of dramatic tableaux totally removed from the little woodcut figures of Dürer or Honter.

Cellarius: The southern sky, 1660. The *Atlas Coelestis* of Andreas Cellarius was an eclectic group of astronomical charts and diagrams, displaying many different and often contradictory views of the heavens. It was partly a historical reference work, explaining the planetary theories of Ptolemy, Copernicus and Tycho. Its elaborately designed celestial maps were essentially artistic variations on a theme. This plate offers a novel view of the Earth (the Pacific and Antarctic regions), as if seen through the starry sphere, from a point in deep space. Undeniably ingenious, its practical use for astronomers is highly doubtful.

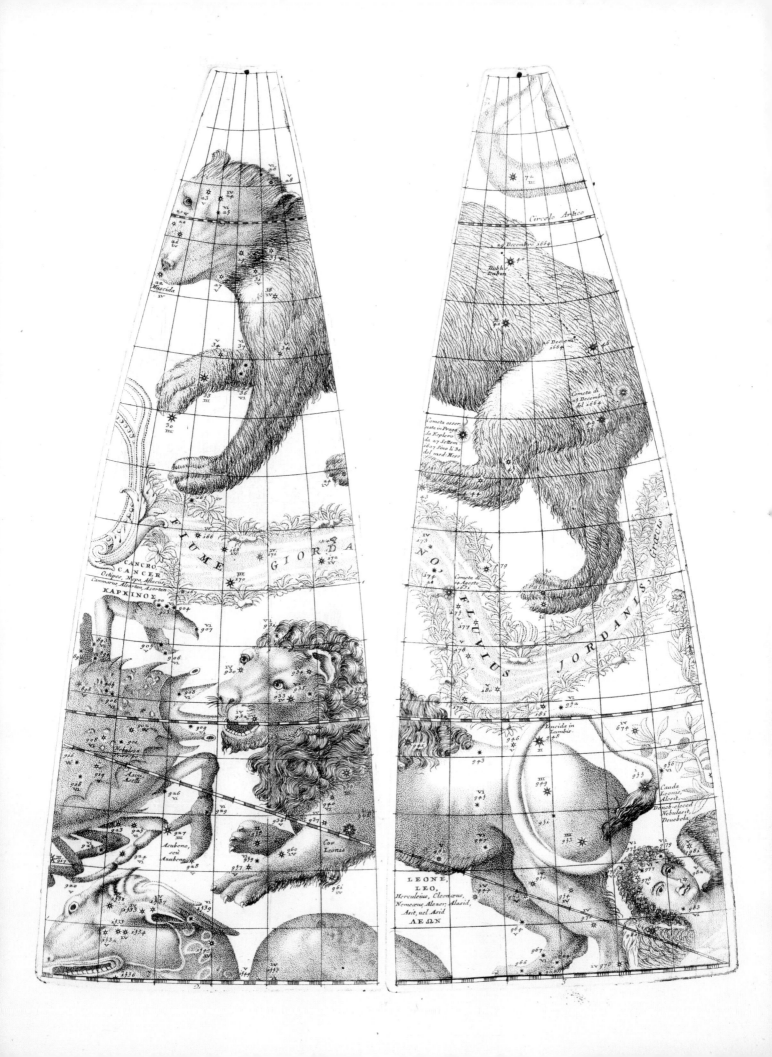

Opposite and right
The Venetian Vincenzo Coronelli brought the craft of globe-making to its height, producing unique hand-painted models for individual patrons, as well as printing globe-gores. These two gores date to *c*. 1701. Although impeccably scientifically based (the inclusion of Arabic names is most unusual) the main object of these richly engraved artefacts was frankly artistic: the positions of the stars are almost lost in the superbly sculptured forms of the constellations.

The publications of Cellarius, Homann and Ottens were hand-coloured after printing, and relied on the rich artistry of the colourist for much of their impact. By contrast, the strength of the Coronelli engravings is most impressive in their pristine state, although they were usually coloured when assembled into globes.

The iconographic style of the constellation figures within star charts does not present a line of clear chronological development, but a number of alternative traditions are discernible. Dürer's highly influential figures are, as one might expect, classically inspired in the main; for example his Perseus is naked like a Greek statue. On the other hand his Orion is armoured like a Teutonic knight of the sixteenth century, while the pointed crown of his Cepheus shows traces of Islamic influence. A few decades later Johannes Honter adopted a thoroughly vernacular style, his male figures dressed for a northern winter like German merchants or courtiers. Elements of this

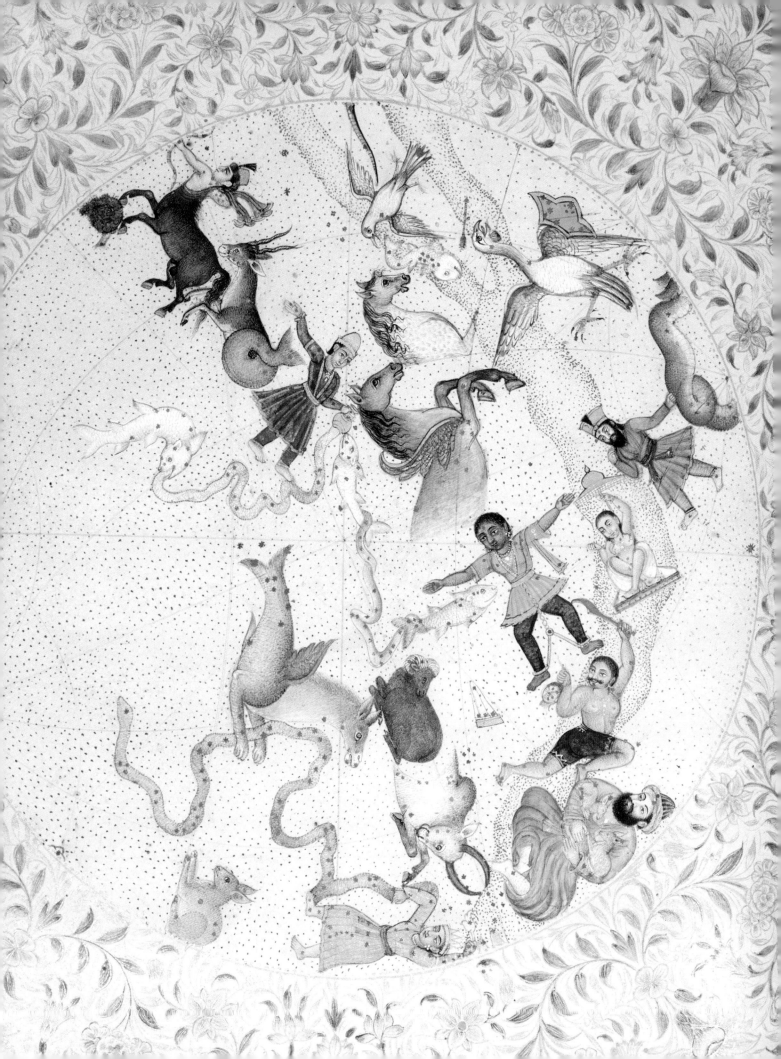

vernacular or contemporary style re-appear throughout the seventeenth
century; Boötes, for example, was always likely to be depicted as a northern
hunter with boots and furs, while Cepheus somehow became cast as an
oriental king rather than a Greek one. But on the whole most mapmakers
aimed at a more or less classical style, the globes of Blaeu and the *Uranometria*
figures of Bayer forming the most widely imitated models. Certain
mythological attributes of the characters which are not strictly derived from
the star patterns tend to be emphasised in this tradition, such as Hercules's
lion-skin, Perseus's winged shoes, or Andromeda's chains. These were
elements which had appeared constantly in the much older pictorial tradition
of the Aratus-Hyginus manuscripts. They were evidently the visual badges
of those characters, and although they were of no scientific significance, the
wider, classically educated public expected to see them. This self-conscious
classicism is not found, of course, in star charts from beyond Europe. Celestial
maps continued to be drawn in Islamic countries down to the nineteenth
century, and artists in India for example had no inhibitions about employing
a lively vernacular style in their constellation figures.

Retrospect: Demystifying the Heavens

The rich array of celestial maps as a published genre between 1500 and 1800
raises the question of why the star chart should have emerged and flourished
so strongly in this period. First there was the historic scientific breakthrough
of the age between Copernicus and Newton, which revolutionised our view
of the cosmos, and which naturally stimulated enormous interest in the
study of the heavens. But on a more practical level there was the impact
of printing, the outstanding cultural event of the era, and the consequent
spread of literacy. Knowledge and skills were no longer transmitted orally,
and any literate person could gain access to the startling discoveries of the
new science. For centuries astronomers, astrologers, calendar-makers and
navigators had received their knowledge of the stars from their teachers,
and in the same way the poet, the physician, the priest or the musician held
in his mind the canons and secrets of his art.

The spread of printing began to change all that, first by offering to anyone
able to read instant access to the accumulated material of centuries in any
field from algebra to zoology; and secondly by fostering the sense that all
knowledge *must* be written down, published and disseminated. The elite,
or arcane, personally transmitted knowledge was becoming a thing of the
past. In the century from 1470 to 1570 all the foundation texts of western
science, philosophy, and literature were taken out of the elite manuscript
collections and offered in print in the market places of Europe. And this
applied not only to texts: at the same time there was a quickening of the
visual imagination, which demanded that the natural and human world be
analysed and presented in diagrams and images. Anatomy, art, architecture,
history, geography and the sciences were all presented in early printed
works enriched with seminal illustrations. This innovation coincided with

Right

Orrery. The first of these mechanical models of the solar system was probably invented under the patronage of Charles Boyle, 4th Earl of Orrery. From a viewpoint outside the solar system they presented the periods of planetary revolution accurately, but the scale of the orbits was necessarily compressed. Some models also showed the planetary moons. Similar models of the Sun-Earth-Moon system were produced, called Tellurians. They seem to embody the eighteenth-century perception of the cosmos as a rationally designed mechanism.

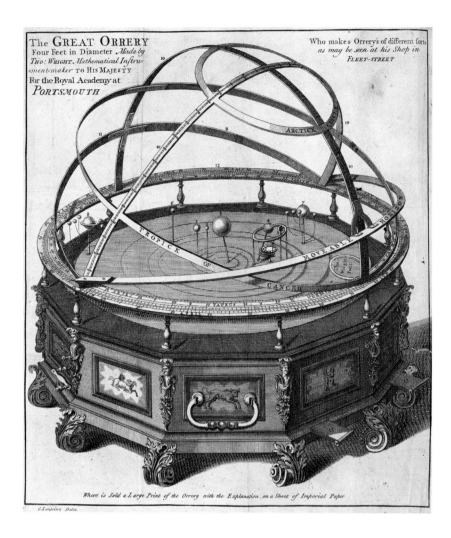

The GREAT ORRERY
Four Feet in Diameter *Made by Tho: Wright, Mathematical Instrument-maker* TO HIS MAJESTY For the Royal Academy at *PORTSMOUTH*

Who makes Orrery's of different sorts as may be seen at his Shop in *FLEET-STREET*

Where is Sold a Large Print of the Orrery with the Explanation, on a Sheet of Imperial Paper

a further force at work in Renaissance thought – the revival of classical models. The overwhelming motive behind the visual art of the sixteenth and seventeenth centuries was to rival or surpass the art of antiquity. The classical constellations were seen as a direct visual link with the science and the imagination of the Greeks and Romans. Their reproduction and dissemination gave an opportunity for educated people to affirm their classical inheritance. The star chart was a small-scale expression of the vogue for classical models, just as the Palladian villa was a large-scale one, and it had the attraction of being linked to an important and fashionable science.

But beneath these facets of cultural history there was a yet deeper, religious, motive which focused the mind on astronomy and its images. This was what later in the eighteenth century came to be called Natural Theology, the attempt to interpret the workings of the cosmos as a guide to divine reason, creative power and purpose. It is striking that many of the greatest astronomers, such as Kepler and Newton, were deeply religious men, consciously seeking to lay bare the laws of cosmic harmony by which, they were convinced, God sustained the universe. One of the earliest texts of this new Natural Theology was William Derham's *Astro-Theology, or a Demonstration of the Being and Attributes of God from a Survey of the Heavens*, which went through

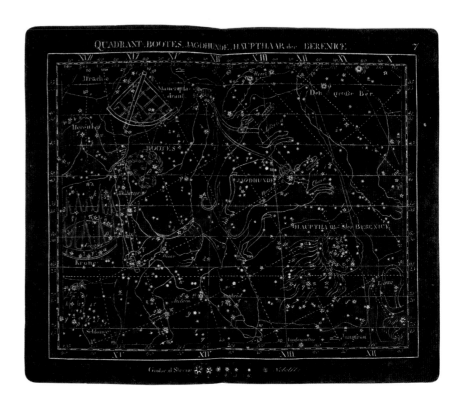

ten editions in the fifty years following its first appearance in 1715. Derham describes the 'new system' of astronomy – that is the Newtonian – as the 'most rational and probable because it is far the most magnificent of any; and worthy of an infinite Creator'. He spends much time speculating that, 'Every fixt star is a sun and encompassed with a system of planets', and that these planets are probably inhabited. The creation of a plurality of worlds is presented as a rational occupation for an omnipotent deity. Derham's entire book is an extended argument for the magnificence of God deduced from that of his creation, and is expressed in purely rational terms. The eclipse of astrology and of theories of celestial influence is total. From the same rationalising impulse as Derham's work came a novel form of solar system map in three dimensions, the Orrery, the ingenious clockwork model of the planets circling the sun. The planets' sizes and distances could not be shown in scale of course, but the periods of their revolutions were, and the more elaborate models included their satellites too. The eighteenth-century delight in celestial mechanics could scarcely have taken a more palpable form than this intellectual toy.

As in the thought of the eighteenth century, so in its images: in the ever-clearer and more complete sky-maps of this period from Bayer to Bode, we seem to detect a process of *demystification* of the heavens, a process of elucidating the riddles of the universe, and placing them within the grasp of rational human thought. This demystification may also be seen in the decline of astrology as a serious science, or as a subject worth considering at all. Demystification appears in one of the most innovative astronomical thinkers of the eighteenth century, Thomas Wright, whose *Original Theory or New Hypothesis of the Universe*, 1750, blended astronomy with a new type

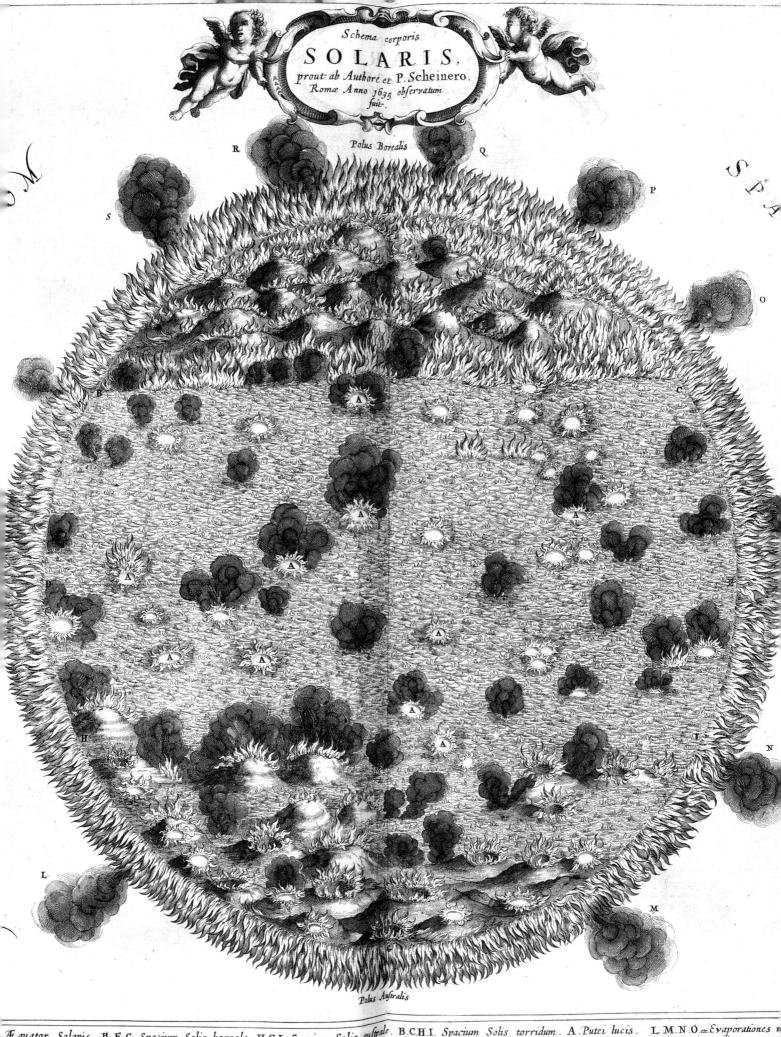

Æquator Solaris . B.F.C. Spacium Solis boreale . H.G.I. Spacium Solis australe . B.C.H.I. Spacium Solis torridum . A. Putei lucis . L.M.N.O. etc. Evaporationes

PLATE XXXII.

Opposite & Above
Thomas Wright: The cosmos, 1750.
From *An Original Theory or New
Hypothesis of the Universe*. Wright,
an instrument-maker and amateur
astronomer, was perhaps the first
man to attempt to rationalise the
large-scale structure of the cosmos.
The Copernican revolution and the
invention of the telescope both
combined to produce a perception
of a much vaster universe. Wright
speculated that it might be infinite,
that the sky visible from the Earth,
with its concentration of stars in
the Milky Way, might be a discrete
system, and that such systems might
be replicated throughout the universe.
Although Wright's ideas were not
empirically based, they were in
some senses prophetic.

of theology in a way that influenced no less a figure than the German philosopher Immanuel Kant. Wright turned from locational star mapping to constructing graphic models of the very structure of the universe, grappling with the distribution of stars, the possible existence of other galaxies, and the idea of a necessary divine centre of the universe. As the vastness of space became evident during the eighteenth century, Wright mused that 'the endless Immensity is an unlimited Plenum of Creations not unlike the known Universe', which was sustained by 'an infinite all-active Power'. Is it too much to suggest that in their desire to chart the heavens, the celestial mapmakers of the seventeenth and eighteenth centuries were actually charting humanity's place within the universe? Confident in the clock-like mechanism of the cosmos, the astronomer sought in the changeless realm of the stars, if it could be measured and understood, a key to the divine mind. By the seventeenth century, the chains of being which had been sought by ancient philosophy and astrology had receded into the past, to be replaced by a scientific quest for the mechanical force which held the universe in equilibrium. The mapping of the heavens, which flourished in a particular form during the formative years of modern science, may be seen as a small but significant expression of a theistic school of natural philosophy. Thus the links between astronomy and religion, observation and divination, which were so strong in the most ancient cultures, and which were so theologically important during the Middle Ages, re-appeared in a new form in the age of science.

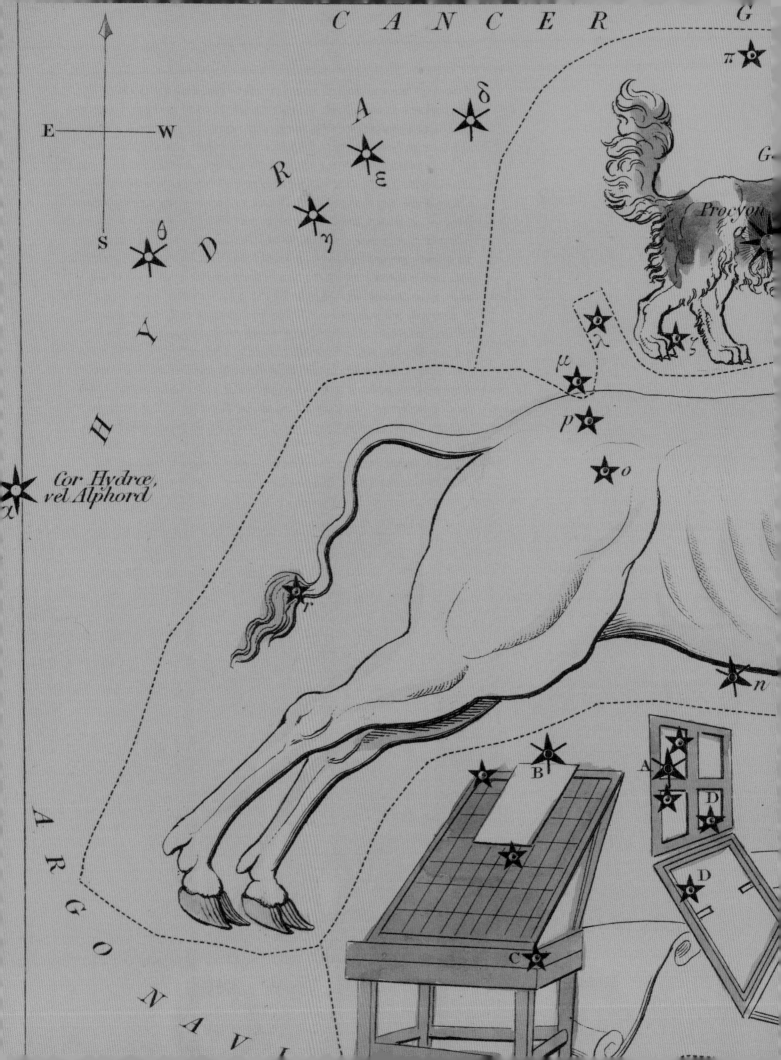

CANCER

G·

π

δ

A

ε

R

γ

δ

D

S

Y

H

Cor Hydræ,
vel Alphord

x

Procyon
α

G·

λ

μ

ρ

o

n

A
C

B
D
C

D

C

A
R
G
O
N
A
V

CHAPTER FOUR

Shifting Horizons

'I feel engulfed in the infinite immensity
of spaces whereof I know nothing and
which know nothing of me ... the eternal
silence of these infinite spaces terrifies me.'
Pascal, *Pensées*, 1657

The Beginnings of Modern Cosmology

When judged by its astronomy, the eighteenth century deserves its reputation as an era of light. Newton's physics was accepted by scientists and theologians as offering a deeply satisfying vision of nature and its creator. Yet the focus of Newtonian celestial mechanics was very much on the solar system. Beyond the paths of the planets was a farther and more mysterious dimension, for the nature of the stars and the structure of the stellar universe were still hidden in mystery. By common consent, modern cosmology began two centuries ago with the work of William Herschel, but despite many startling breakthroughs, no second Newton has appeared to flood our eyes with light. On the contrary we seem to advance through a tortuous maze in which no map can aid us. In this period of astronomy, a new visual dimension has been added through space photography, and the *mapping* of the heavens has arguably given way to direct *images* of the heavens, from even the farthest reaches of space. In the shaping of modern cosmology, the interaction of thought and image has had enormous consequences for the professional astronomer, and has assumed a dominant role in the popular mind.

William Herschel (1738–1822) achieved fame as the discoverer of the planet Uranus in 1781, becoming the first man since the dawn of classical astronomy to enlarge the bounds of the solar system. His discovery came in the course of the monumental new sky survey in which he was engaged, with no less an object than to lay bare the structure of the universe, to see beyond the flat surface of the starry sphere and grasp its three-dimensional form. The locations of the stars as observed by Ptolemy or Tycho or Flamsteed were expressed purely in terms of angular measurement, and this information lay behind the traditional two-dimensional star chart. Even the celestial globe, although it is a three-dimensional object, actually shows the sky only as a surface, in two dimensions. By the late eighteenth century the desire to penetrate beyond the shell of the starry sphere and gauge the depth of the universe had become urgent. The general view of cosmology in this period, with all its uncertainties, was admirably summed up in 1780 by Samuel Dunn, a writer on astronomy and geography, who annotated his star chart:

> The number of stars which have been accurately observed is no more than 3,000. Yet by the help of telescopes, 21 stars have been seen in the space which forms the cloudy star [i.e. the nebula] of Orion's sword, 36 in that of the cloudy star of Pegasus, 78 in the asterism of the Pleiades, and 2,000 in the constellation of Orion. From which it has been conjectured that the number of fixed stars is no less than 10 million, beside those of the Via Lactea, or Milky Way, and such as cannot be discerned by the best glasses, and that twice or thrice that number

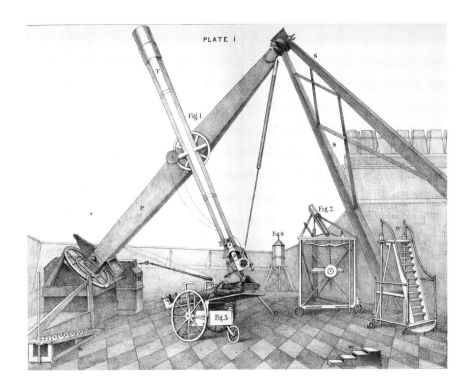

Right
A state-of-the-art telescope of
the mid-nineteenth century; from
Henry Lawson's *A Paper on the
Arrangement of an Observatory
for Practical Astronomy*, 1844.

Above
Model of William Herschel's telescope.

would cover the expanse of Heaven. The bright star Syrius in the constellation of the Great Dog hath been estimated to be distant from us more than 2 million of million miles. And the distance of a star of lesser apparent magnitude hath been found to be more than 30 million of million miles. Wherefore tis concluded that every fixed star is a sun, having planets and comets moving round it like those which move round our sun. From all of which it may be concluded that, if the universe doth not extend itself beyond the powers of number, weight and measure, it may extend too far for human reason to comprehend. Particles of light are inconceivably small, hard bodies thrown off from the sun, stars etc. which move at the rate of 10 million miles in a minute of time and come from the sun to our earth in 8 minutes, and from the fixed stars in about six years.

It is clear from writers such as Dunn that the problems of cosmological scale were beginning to be understood, even if the answers were remote and mysterious. Herschel's novel self-imposed task was to estimate the radial distances of stars from the Earth. In the classical doctrine of the starry sphere, the stars were conceived to be each of differing brilliance, but all set in one spherical shell located at a specific distance from the Earth. This doctrine was weakened and finally shattered by a number of blows. First, in the Copernican system the revolution of the Earth should have revealed obvious stellar parallaxes; that it failed to do so suggested the possibility that the stars were at a far greater distance than had been imagined. Second, the actual existence of the starry sphere became impossible in the work of Tycho, Kepler and their successors. Third, the telescope revealed in any given field of view more stars than had been visible to the naked eye, and the more powerful

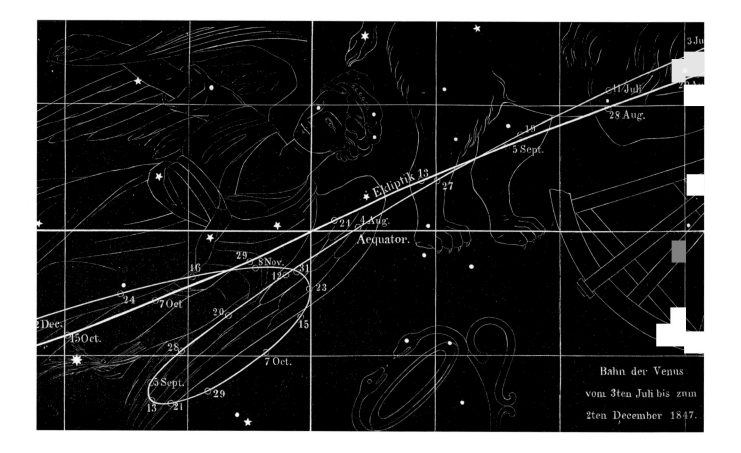

On the map, the following labels appear: 3 Ju, 11 Juli, 28 Aug., 5 Sept., 19, Ekliptik 13, 27, 21, 4 Aug., Aequator., 29, 8 Nov., 31, 16, 13, 23, 24, 7 Oct., 20, 15, 2 Dec., 15 Oct., 28, 7 Oct., 5 Sept., 29, 13, 21, Bahn der Venus vom 3ten Juli bis zum 2ten December 1847.

Above

Johann Müller: Path of the planet Venus, 1847. A long-standing problem with star charts was their unsuitability to show planetary and lunar positions. Because they are in motion against the background of stars, their positions would necessarily appear as lines which would tend to dominate the map, as we see here, while their irregular paths means that any map would be valid for a few months only – a decisive factor for the commercial map publisher. Planetary positions were published in textual form as 'ephemerides'. Müller's map is an interesting example map of a planet's position made for purely demonstrational purposes in a scientific textbook.

the instrument the more stars appeared. Finally the very notion of the 'fixed stars' was challenged in 1718 by Edmond Halley's announcement that he had detected small movements ('proper motions') among the stars. Subsequently this movement of the stars was found to be caused partly by the motion through space of the solar system itself, but also to be shared by all stars in their own right. One incidental effect of these movements will eventually be to destroy the constellations as we know them: in a quarter of a million years they will be clearly re-shaped, and in half a million years they will be unrecognisable. All these discoveries had the cumulative effect of suggesting strongly that the stars were *scattered* throughout space at hugely varying distances from the Earth. In this view the different magnitudes of brightness were explicable simply in terms of distance from the Earth. But given that this was so, was there a structure to the realm of the stars, or was the distribution random? Galileo and the other early telescope users had resolved much of the Milky Way into stars: what was the significance of this concentration of stars?

To decide these questions, to map the universe in three dimensions, it was obviously necessary to know stellar distances; but in the absence of stellar parallaxes, this was precisely what Herschel did not know. He overcame the difficulty by making the assumption that all stars are of similar absolute brilliance and that stellar magnitude was therefore an index of distance from Earth. This was a very big assumption, but according to the law of averages it was not wildly inaccurate. Working on this simple principle, Herschel proposed that the Milky Way held the key to the structure of the universe:

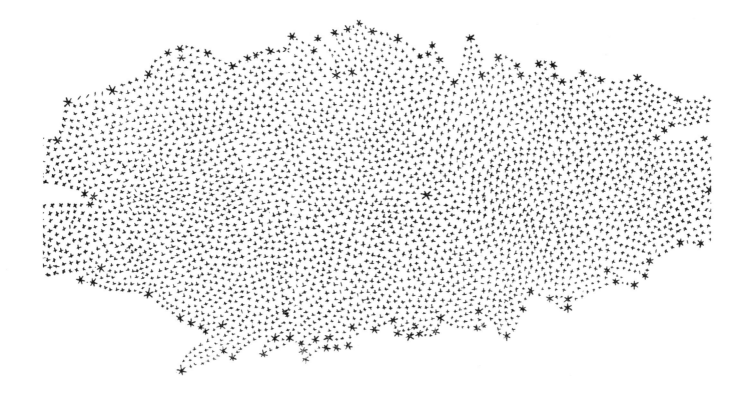

the universe was a disc-shaped mass of stars which we see edge-on as the
Milky Way. When we look at right angles to the Milky Way we are seeing
through the thinning edges of the disc towards empty space. The tentative
model which he published in 1785 shows the Sun near the centre of a rather
ragged disc. The earlier work of Thomas Wright was purely hypothetical,
but this drawing by Herschel marks the dawn of observational cosmology.
In fact, however, as he built and used more and more powerful instruments
(the largest was a 40-foot reflector), his faith in his early theory was shaken,
for he perceived that certain of the nebulae, the clouds resembling luminous
dust that were far removed from the plane of the Milky Way, were resolvable
into myriads of innumerable stars; on his disc-model there should be no high
star concentrations at those places. Herschel undertook what was the most
ambitious and detailed sky survey made to that date, which involved the
counting of hundreds of thousands of stars subdivided within minute fields
of view, in order to study patterns of star distribution and concentration.
This work did not, however, lead to new published star charts, since that
was not its purpose. The problem of understanding the scale of what the
observer was seeing would haunt astronomy for a further century and a half.
Nor was there yet any key to the internal nature of the stars, for the science
of astrophysics had not yet been born. Even a man like Herschel could not
envisage how astronomers could ever obtain knowledge of the processes at
work in the Sun, and he was capable of speculating that the Sun was a 'lucid
planet', possibly inhabited by men who were protected by thick clouds from

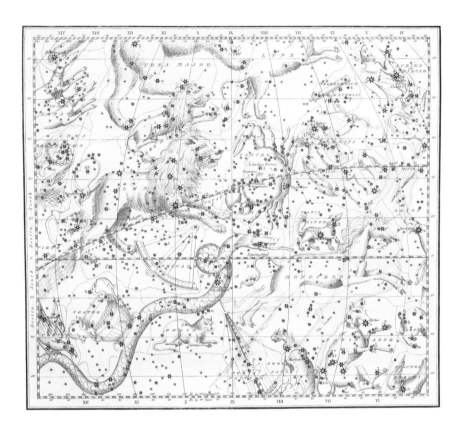

Right & Below

James Middleton, *Celestial Atlas*, 1843. By the nineteenth century, pictorial star maps had ceased to be used by serious astronomers, but they were still published for the popular market. Middleton's elegant engravings show the constellations in their correct orientation as seen from the Earth. They were accompanied by plain charts without the figures to aid observation. The canon of the constellations was by no means fixed at this date: the Telescope shown here had been proposed in honour of Herschel, but it did not become permanent – the southern Telescope is a separate star group. The small cat south of Hydra was another short-lived astronomical caprice.

its fiery upper atmosphere. The technical advances in chemistry, physics and instrumentation that would bring astronomy into the laboratory still lay in the future, and such speculations were commonplace even among leading scientists.

Similarly in terms of star mapping, the first half of the nineteenth century did not see any radical break with the past. It is often claimed that Bode's great celestial atlas of 1801 marked the end of the pictorial star chart tradition, but this is not accurate. It is true that no pictorial star atlases more detailed or more elaborate than Bode's were produced, but smaller atlases very much derived from Bode continued to appear throughout the century, some frankly popular but others claiming to be based on original observations, and offering new, practical chart design. It became fashionable to publish star maps, especially monthly or seasonal ones, white on black, to match the appearance of the night sky. When a recognisable skyline – that of London or Paris for example – was drawn at the base of the map, the distinctly artistic illusion was achieved that one was surveying from the comfort of one's study the night sky over Greenwich or Montmartre. James Middleton's *Celestial Atlas* of 1843 combined austere white on black seasonal maps with some elegant pictorial charts in the traditional style. A more playful novelty was Samuel Leigh's *Urania's Mirror*, 1823, consisting of a series of constellations printed on cards with pierced holes marking the principal stars; when held up to the lamplight the star patterns sprang clearly into shape.

Leigh's novel innovation was not widely copied. But a new type of popular star map that appeared in the mid-nineteenth century did become

Above
The revolving planisphere is the most practical form of star chart ever designed: by masking off part of a map of the entire hemisphere, it is able to represent sections of the sky actually visible to the observer. The hours of the night and the days of the year are calibrated around the edge of the disc, so that its rotation mirrors that of the Earth itself. It has affinities with the clock face and with the astrolabe, from which in turn the clock had evolved.

Right & Next Spread
Samuel Leigh, *Urania's Mirror*, 1823. An ingenious novelty typical of its time, these were cards showing the constellations, pierced with holes to mark the star positions. When the card was held against a light, the constellation appeared as in the night sky, an excellent demonstrational or mnemonic device. The printing equipment illustrated here below Monoceros and Canis Minor was one of a number of short-lived constellations with technical subjects proposed by early nineteenth-century astronomers (another was Montgolfier's balloon), mainly to fill areas in the southern sky.

permanent – the rotating planisphere. Whether consciously modelled on the astrolabe or not, the planisphere works in a similar way by separating the locating map of the stars from its frame of reference, but with this difference – that it is the time of visibility that is separate and movable. Basically it is a projected map of the northern or southern heavens centred on the pole. By masking approximately half of the map, the planisphere shows that part of the sky which is actually visible at any given moment. Like the astrolabe, the planisphere is valid only for a given latitude.

Astronomy was promoted in the nineteenth century as a distinctly moral branch of science, one which awakened the mind to religious perception. James Middleton summed up this attitude when he wrote in the preface of his 1843 collection of star charts:

> The man of science may be enabled to discover in the more distant parts of the universe, the same laws and regularities which govern our own system; while the Christian will derive from the science which it teaches, the most sublime illustrations of the wisdom, power and majesty of that Being who garnished the heavens, who telleth the number of the stars, and calleth them all by name.

The perceived conflict between science and religion was to centre on geology and biology, while astronomy, however challenging its findings might be, always inhabited an ethereal realm which was not felt to be at odds with a religious sensibility.

Far removed from popular astronomy, however, a new style of scientific star map was taking shape. In the first place, the research scientist no longer had any need of the traditional constellation figures, which were a distraction on a serious map. But more important, even the best of the traditional star atlases – that of Bode – showed only some 15,000 stars, the vast majority

α

Capella, vel Alioth

E————W

S

γ

ε

η

Duo Haedi ζ

A U R I G A

b ι

μ

Algenib

c δ ζ σ

ν

f ε

Menchib ε ς

ξ ν

β να

ω

Atik

MU

CASSIOPEIA

Schedir
α

k

γ

π

o

ssi

Almaach
γ

ν

μ

A N D R O M E D A

p

β
TRIANGULA
γ

Mirach
β

C A

α

Sid.ʸ Hall sculp.ᵗ

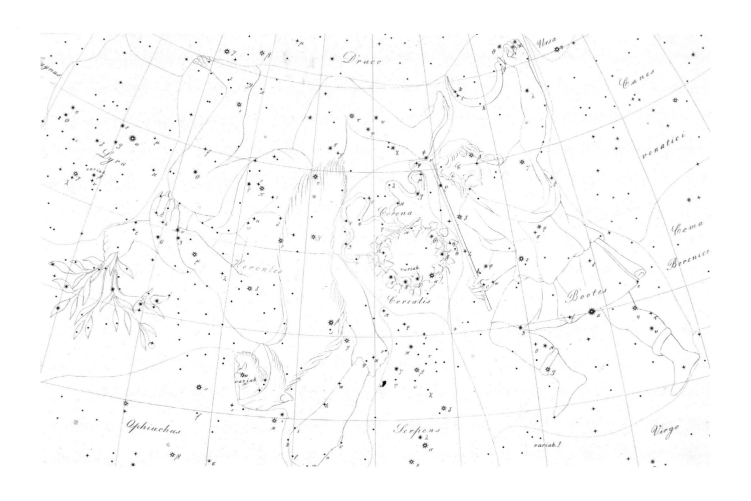

Opposite
In 1843 Friedrich Argelander
published *Uranometria Nova*,
an unrevolutionary celestial atlas,
showing the traditional constellation
figures and tentative boundaries
between constellation-sectors
(Opposite Above). He was soon to
begin work on an epoch-making
sky survey resulting in a new type of
star map for a new age of rigorous
astronomy. The *Atlas des Nordlichen
Gestirnten Himmels* of 1863 showed
no constellations, no boundaries,
no names, no symbols of any kind,
merely an austere monochrome
fabric composed of 324,189 stars
(Opposite Below). Even though it
reached initially only to −2 degrees,
each page showed many thousands
of stars. Based entirely on optical
sightings, Argelander's work was
the last monument of the pre-
photographic age of star mapping.

Next Spread
A. P. Herbert: *A Better Sky*, 1944.
Herbert started from the premise that
we know or care little for astronomy
and the stars because their mythology
and their names are alien to us. He
proposed a reformed, rational sky
rooted in modern experience. He
renamed the constellations and all
the principal stars. Ursa Major became
Great Britain, with Shakespeare, Wren
and Johnson as the stars; Draco
became the Tyrants, composed of
Robespierre, Hitler and Mussolini;
Cassiopeia was transformed into
the United States with Washington,
Jefferson and Lincoln; Orion, with the
stars Columbus, Cook and Nelson,
became the Sailor. Sadly for Herbert,
his reformed astronomy shared the
fate of all other similar projects.

being naked-eye stars, while the astronomer of the mid-nineteenth century was handling telescopic observations of ten times that number. This situation was comparable perhaps to a pioneer Alpinist planning his climbs with the aid of a map of the whole of Europe. To rectify this situation Friedrich Argelander published in Bonn his *Atlas des Nordlichen Gestirnten Himmels*, 1863, commonly referred to as the 'Bonner Durchmusterung' (Bonn Survey), which showed a staggering total of 324,189 stars. As a work of meticulous observation and draftsmanship, it commands our amazed admiration, tempered perhaps with regret that its austere cartography represents the final act in the demystification of the heavens. Initially covering only the northern hemisphere, it was extended by Argelander himself to a declination of −23 degrees by 1886, and by a different team working in Argentina to −62 degrees in 1908, finally mapping the south polar stars in 1930. The accompanying Bonner Durchmusterung star catalogue was for over half a century the internationally recognised celestial reference system. The Bonner Durchmusterung atlas shows no constellations. Argelander had earlier published a very different atlas, the *Uranometria Nova* of 1843, one of whose aims was to define a fixed canon of the constellations and to introduce recognised boundary lines between them. Serious astronomers were seeking to rationalise and clarify the map of the heavens; one of the last innovations among the constellations was the division of the very large southern Ship into four: Vela the Sail, Pyxis the Compass, Puppis the Stern, and Carina the Keel. In practice there was no change to the constellations after the middle of the nineteenth century, although the definitive modern list of eighty-eight was not officially endorsed until 1930 under the authority of the International Astronomical Union. This has not, of course, prevented occasional polemical suggestions for a new sky order, such as that of A. P. Herbert. Herbert's rationalised map of the heavens is perhaps the kind of scheme which the revolutionary government of France might have promoted in the 1790s following its rational calendar reform, if it had not been too busy executing its scientists.

The Photographic Revolution

Argelander's work was particularly impressive for its base in telescopic observation and its use of conventional plotting. Yet it coincided exactly with the first astronomical use of photography, which was to revolutionise many of the methods and the fundamental concepts of astronomy. Beginning in the 1850s the Moon and the Sun were naturally the first objects to be photographed, followed by the most photogenic planets, Saturn and Jupiter. It was inevitable that the new technique would make an enormous impact on star mapping, for two principal reasons: the time-exposed photographic plate could detect light-sources far fainter than the human eye, and it could preserve the image for later study and analysis. In 1887 a conference in Paris initiated a vast international astro-photography programme with the aim of publishing a new *Carte du Ciel*, which would be an atlas not of redrawn

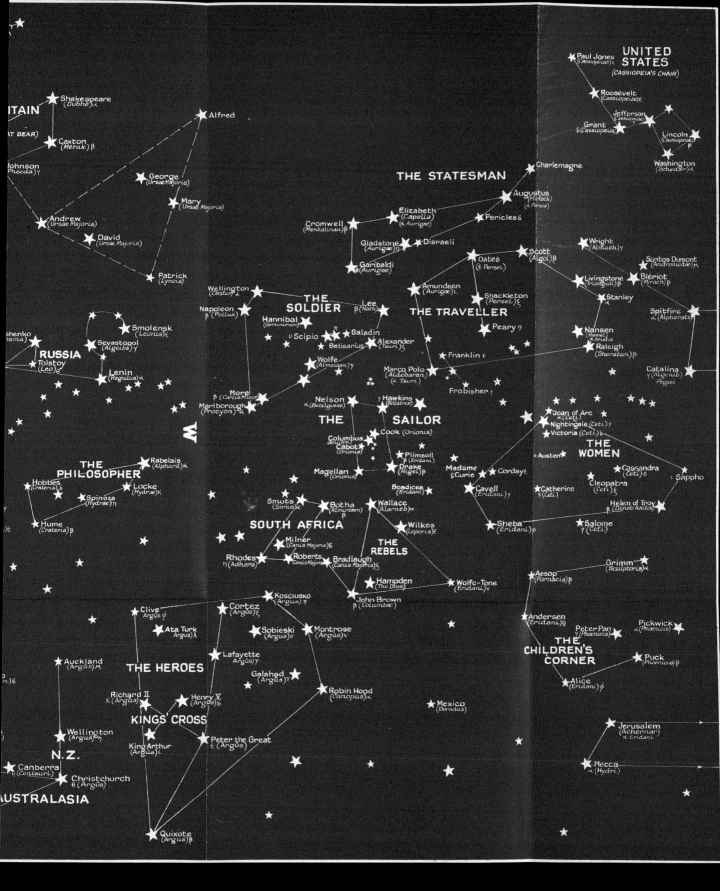

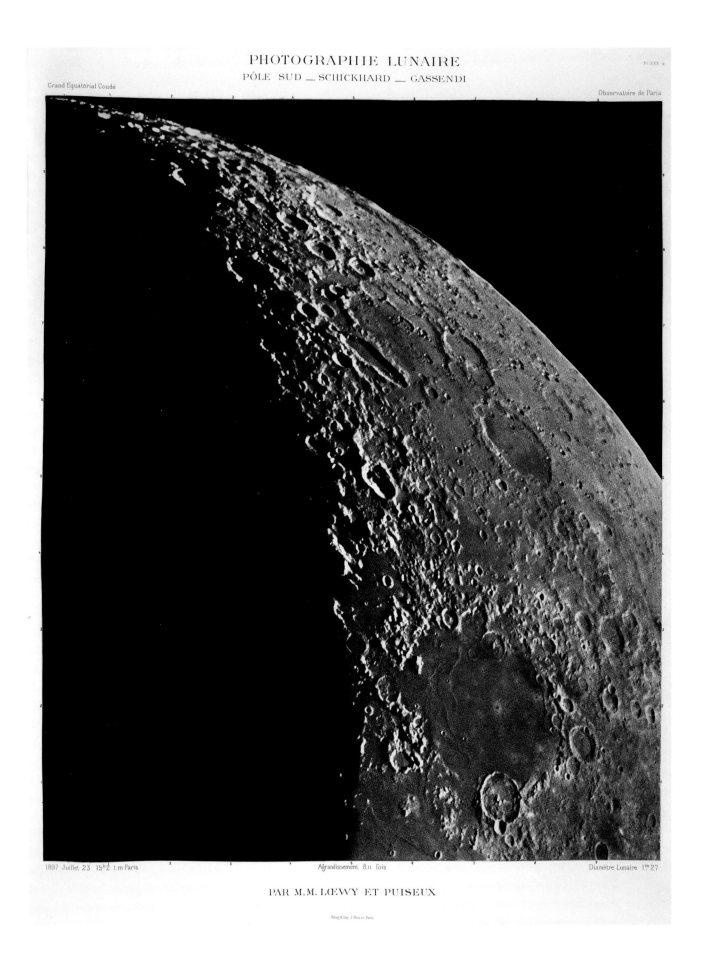

Grand Equatorial Coudé

Observatoire de Paris

1897 Juillet 23 15ʰ2 t.m.Paris   Agrandissement 8.11 fois   Diamètre Lunaire 1ᵐ 27

PAR M.M. LŒWY ET PUISEUX

Opposite
Charles Le Morvan, The Moon
with Schickhard and Gassendi
Craters, 23 July 1897. From
Maurice Loewy and Pierre Henri
Puiseux, *Atlas photographique de
la lune*, 1896–1910.

Below
Photographic negative print
taken with the 48 inch Samuel
Oschin telescope at Palomar
Observatory. Photograph taken
11/12 August 1950.

maps, but of photographic prints, complete with added co-ordinates. Very much in the organisers' minds were the recent pictures by the French brothers Prosper and Paul Henry of the Pleiades, which showed more than 1,000 stars compared with less than 100 detectable even with a good telescope. Owing to organisational problems, this programme dragged on for 80 years, while other more tightly focused projects overtook it. One such, which resulted not in published maps but in a vast catalogue of the southern stars, was the *Cape Photographic Durchmusterung* which Jacob Kapteyn produced between 1896 and 1900. Working in a laboratory in Groningen, Kapteyn used a series of photographic plates from the Cape Observatory to measure the co-ordinates of 454,875 stars between declination –18 and the south pole, down to the tenth magnitude. This desire for absolute precision and completeness was of course an ever-receding mirage, since improving astro-photography would always bring more and more stars into view. The summit of this type of photography was reached with the National Geographic–Palomar Observatory Sky Survey of 1954–58. A series of 1,758 plates, each showing thousands of stars reaching down to magnitude 21, it expanded the published image of the universe by twenty-five times. It carried the observer across a billion light years of space and its stars are as uncountable as those in the sky itself.

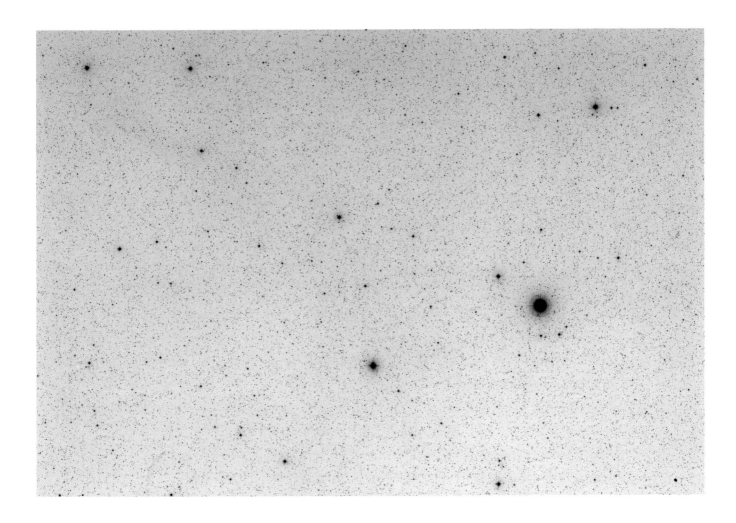

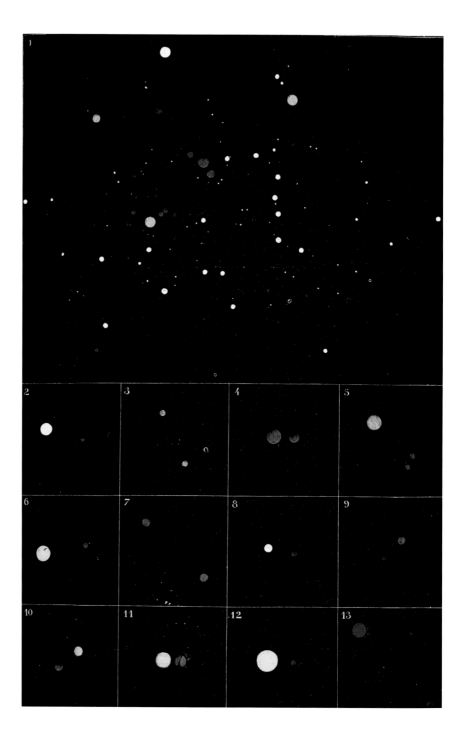

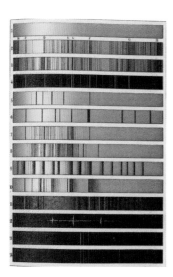

Scarcely less important was the new technique of spectroscopy, developed
in 1860 by Gustav Kirchhoff and Robert Bunsen. It had always been thought
that no direct knowledge of the composition of the Sun and stars would ever
be possible. But the discovery that the analysis of light could reveal in detail
the chemical and physical characters of the light source laid the foundation
of astrophysics, and later of understanding the processes of evolution in
the universe. Working with stellar spectra just a few years later in 1868, the
English astronomer William Huggins made a discovery that was to assume
the first importance in modern cosmology. Analysing the light emitted by

the star Sirius, he found that its spectrum was shifting markedly to the red.
This phenomenon had been theoretically described in general terms by the
Austrian physicist Christian Doppler, as an effect of the changing wavelength
of a moving energy source – it applies to sound as well as light. Huggins's
calculations indicated that Sirius was moving away from the Earth, at a
velocity of around 100,000 mph. The full implications of this effect would
become clearer half a century later, when related findings would be gathered
from distant galaxies.

By accident, the discovery of spectroscopy coincided with the invention
of colour printing in the 1860s. It had always been known that stars shone
with many different colours – Ptolemy had spoken of 'golden-red Arcturus'
– but these colours could now be printed and offered to a wide public. In
time their scientific significance would become clearer, and it was in the
mid-1950s that the Czech astronomer Antonin Becvar pioneered the use
of colour and symbol in published star maps to indicate a wide variety of
analysed information about the stars. The star atlases of today, both those
for the professional and for the amateur sky-watcher, do more than locate
positions, for they are able to draw on the results of spectroscopy and
photography. They may show the spectral class and magnitude of the stars,
and whether they are multiple or variable, as well as non-stellar objects such
as galaxies, nebulae, clusters and gas-clouds. The mapping of the heavens
has now progressed beyond the locational, which has become the province of
photography, while redrawn maps may be used to display levels of thematic
information derived from wider analysis.

The excitement of photography was just one of the reasons behind
a new genre of popular astronomy books which proliferated in the later
nineteenth century. Proctor and Dunkin in England, and Guillemin and

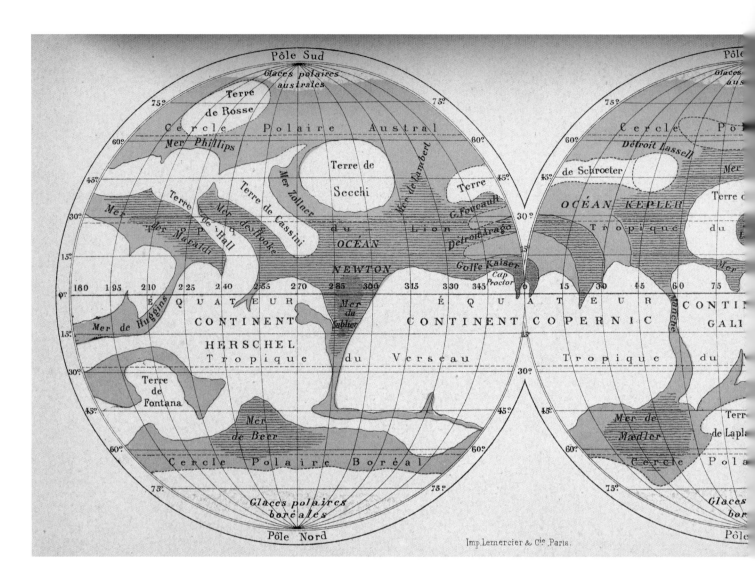

Previous Spread
Guillemin: The night sky over Paris, 1865. Guillemin was probably the inspiration behind Dunkin's similar maps of the sky over the London skyline. Guillemin has drawn the architecture of Paris with great care to create an illusionistic effect.

Above
Map of Mars from *Astronomie populaire* by Camille Flammarion, Paris, 1880.

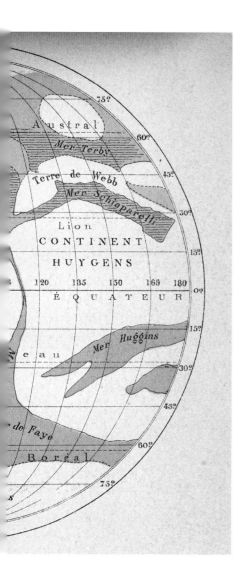

Plate 15.

Above, right
Isaac Roberts: four spiral nebulae
(M.64 Comae Berenicis, M.63 Canum
Venaticorum, Spiral Nebula H.iv 76
Cephei, M.51 Canum Venaticorume).
In the late nineteenth century
astrophotography revolutionised
astronomy in two ways: a time-exposed
film could capture more light-sources
than the human eye even when the eye
was aided by a telescope; and it could
preserve the image for future analysis.
Isaac Roberts was one of a number of
amateur enthusiasts who pioneered
astro-photography, in particular in
devising the essential clock-driven
mounting which maintained the
camera's field of view. These pictures
were exposed for between 90 minutes
and two hours and fifty-six minutes.
The study of nebula photographs was
to assume ever-greater importance in
modern cosmology.

Flammarion in France, produced summaries of the latest findings illustrated with pioneering photographs of eclipses, of the rings of Saturn, and, before the end of the century, of the nebulae that were increasingly engaging the attention of astronomers. In this period the dividing line between what could be achieved by professional and amateur astronomers was by no means fixed, and landmark discoveries and photographs could still be made by amateurs such as Norman Lockyer, who identified helium in the sun, and Henry Draper in New York, who took the first photograph of the Orion nebula in 1880. The many high-quality photographs by Isaac Roberts became features in astronomical books and journals between 1890 and 1920. Part of the excitement of astro-photography was undoubtedly the ambiguity of some of the new features it revealed. Intense interest in the possibility of life on Mars was aroused by the interpretative drawings based on photographs by Percival Lowell and Giovanni Schiaparelli. The books of Camille Flammarion in particular, one of the great nineteenth-century popularisers, are a heady mixture of science and imagination, photography and fantasy.

Photography and spectroscopy placed powerful new tools in the hands of astronomers, and the secular development of physics and chemistry prepared the ground for a new phase of modern cosmology, in which the problems of scale and structure in the universe were determinedly addressed. It was the interpretation of nebulae which proved to be decisive. Some of the great nebulae are visible to the naked eye and Ptolemy had described half a dozen, including the Andromeda nebula, as 'misty, cloud-like stars'. Their luminosity was no different from that of a normal star, and on the luminosity-distance principle they had to be part of the familiar star system, just as Sirius or Antares or any other star was, although they clearly differed in some way from those other stars. Herschel's observations with his largest telescope, however, raised the problem that some nebulae appeared to be actually composed of stars. This was an isolated problem to which there was no answer, and even in 1885 the Andromeda nebula was thought to be some type of nova remnant, situated perhaps 30,000 light years away – a vast enough distance but well within the bounds of our star system.

The real difficulty came in the early twentieth century when ever-improving instruments and photographs resolved this nebula and most others into what were unmistakably discrete systems of stars in their own right. It was Edwin Hubble who, working from the known properties of certain variable stars found in the Andromeda nebula, was able to calculate its distance as approaching a breathtaking one million light years (and this figure was subsequently doubled). The conclusion became irresistible that this was a star system paralleling our own Milky Way, and vastly distanced from it. It was at this stage, the 1920s, that the word galaxy assumed its modern meaning. The word nebula has remained in use, and some of them are indeed clouds of gas; but the majority are more correctly described as galaxies, vast but discreet star groups like our own, comparable to islands in the ocean of space.

Contemporary Cosmology and its Images

One of the great turning-points in modern cosmology was Edwin Hubble's confirmation of certain earlier findings, namely that the light from these distant nebulae is shifted decisively to the red end of the spectrum, which is the characteristic sign of a receding energy source. Hubble was able to show that this recessional velocity was proportional to the distance from the Earth: the further away these galaxies were, the faster they were travelling. The velocity of these distant galaxies was staggering, measured in thousands of miles per second, and such recession was observable in every part of the universe.

The scale of the universe, and its state of dynamic growth, have been discoveries as revolutionary and disturbing to the modern mind as the Copernican theory was in the sixteenth century. Their implications seem to take us to the limits of human thought, not least in the way that they fuse space and time together. As early as 1676 the Danish astronomer Ole Rømer had demonstrated that light travels at a finite speed, but the full implications of that fact in the field of cosmology emerged only as the true scale of the universe came to be appreciated. Modern instruments can now 'see' some fifteen billion light years into space. But of course what is seen is the received light which left its source fifteen billion years ago. So in looking across the universe like this, we see a cross-section of time, of the history of the universe, as well as space. There is no limit to the universe, only a 'cosmic event horizon' which we can never cross because light itself defines it, and light can never be overtaken. It is no accident that the age of the universe is also conceived to be of the order of fifteen billion years, since that is the time the universe has taken to arrive at its perceived extent: whether the scale is one of time or space, the figure is the same. Quasars, for example, are observed only in the most distant regions of the universe; they are considered to have evolved comparatively soon after the birth of the universe, and to be the nucleus of some types of galaxy. Hence the correlation between time and distance is clear: in looking across the universe to distant quasars we are looking back to embryonic galaxies which no longer exist in that form. We are invited to conceive of a universe that is in some way finite in space and time, but whose extent can never be measured even in theory. We may feel free to doubt whether these paradoxes are truly objective, physical realities, or will prove to be categories of human thought and perception. The paradox of the universe's endless growth through unimaginable levels of energy change is captured in the evocative phrase 'catastrophic evolution'.

Right
Cosmic microwave background radiation. Data collected from the Cosmic Background Explorer satellite shows temperature fluctuations in cosmic radiation. This is claimed to relate to the uneven distribution of matter in the universe, and to be comparable to physical ripples or echoes of the Big Bang with which the universe began.

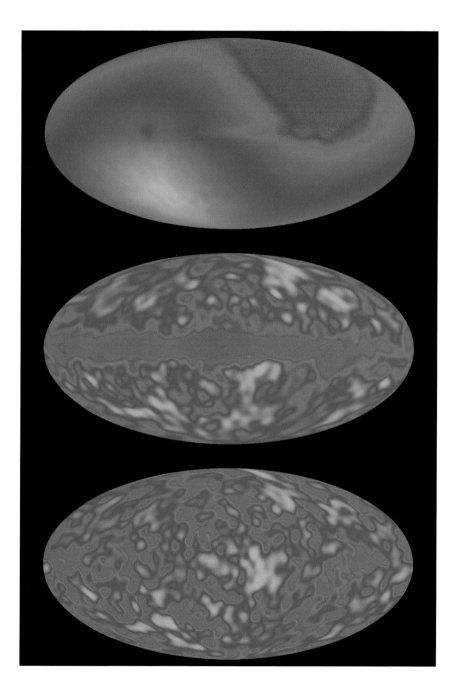

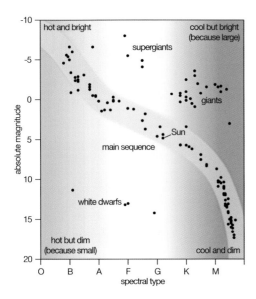

Modern cosmology is now driven more by mathematics and theoretical physics than it is by traditional observation. Cosmologists build mathematical models of what is happening in stars, in galaxies and in the universe as a whole, and by extending these models they hope to discover how the universe arose, how it has reached its present state, and how it may develop, and even end. Some, perhaps most, of these mathematical models are incomprehensible to lay people when they try to translate them into plain language and observable physical reality. Hence arise concepts like curved space-time, black holes, neutron stars, dark matter, and the quantum universe or multiverse. In classical astronomy, observation preceded conceptualisation. Now it seems that the mathematical model comes first, and then the astronomer seeks for empirical confirmation, as in the famous case when theorists of the Big Bang predicted that the afterglow radiation of that event should still be present, and it was indeed found.

The contemporary perception of a universe of transcendent mystery has been fed by a new generation of dramatic astronomical photographs whose appeal is both aesthetic and intellectual. We now see across billions of miles of space to multicoloured nebulae, galaxies, interstellar dust and supernovae, as an earlier generation once delighted in images of sunrise, storm-clouds and mountains. As celestial maps have become ever-more austere and functional, so the art of the astro-photographer has become more lavish and kaleidoscopic; the taste for visual drama which was once satisfied by the mythology of the constellations has been transferred from maps to photographs, and the results are often breathtaking. They represent worlds the human eye will never see, whose reality we take on trust from scientists, as once descriptions of heaven were taken on trust from prophets or priests. These images have undoubtedly been used to reinforce a new secular mysticism that seems to surround current cosmology, purveyed in works with titles such as 'Physics and the Mind of God', or 'The Afterglow of Creation', or 'The Search for the Infinite'. From the seventeenth century onwards, Newtonian physics was widely seen as promoting a mechanistic view of the universe in which the image of the clock was used repeatedly. This was not irreligious but was used to underpin a Christianity that was consciously rational and self-satisfied. The mapmakers of this period were undoubtedly part of this process of demystification. The twentieth-century realisation of the unimagined vastness and dynamic movement of the universe has re-awakened a profound sense of mystery, almost of dislocation, in human thought. The placing of astronomy within a religious framework is, as this book has tried to show, a deep-rooted tradition. But it is perhaps surprising that in this secular age, the language of physics should reach back to and revive the language of mysticism. In popularising this kind of thinking, the map of the heavens has been replaced by the image of the heavens, presented in the forms which our age demands.

Perhaps the nearest direct approach to metaphysics in modern astronomy is the discussion of an 'anthropic principle': the fact that the universe is, at any level, comprehensible to the human intellect is thought to indicate a special affinity between humanity and the cosmos. In its extreme form

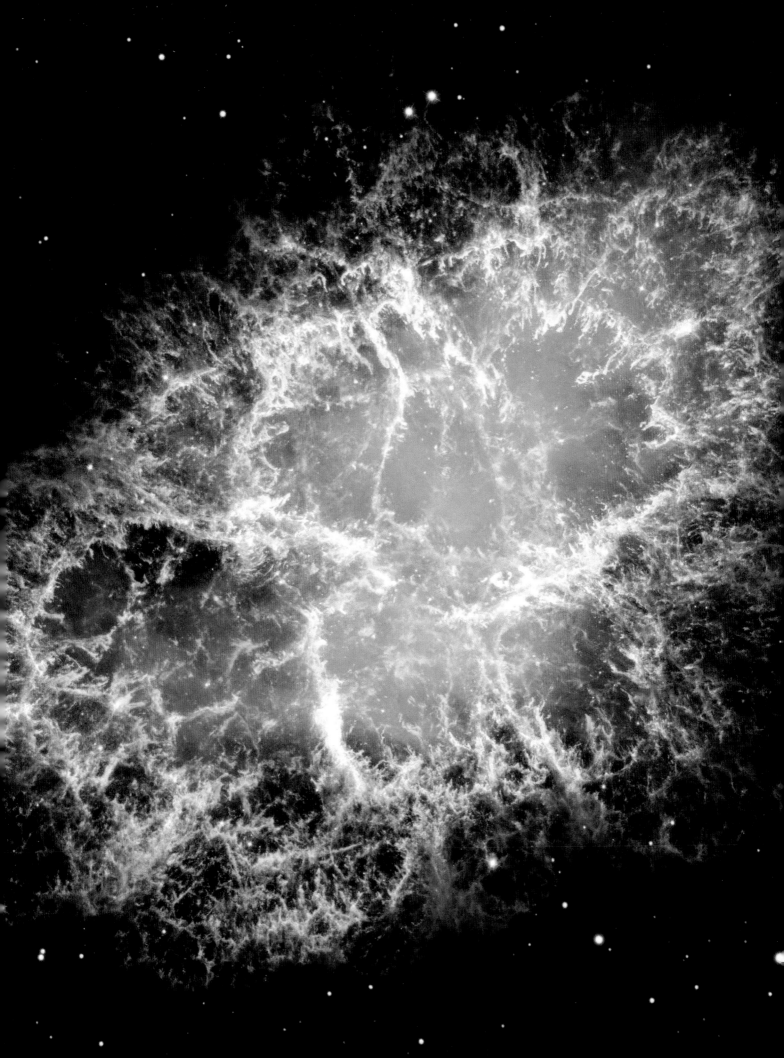

the anthropic principle even suggests that the purpose of the universe is somehow connected with humanity, and that humans have a special role in the evolution of the universe because we are – as far as we know – the only part of the universe which understands the rest of it. There is an intriguing circularity about this: the emergence of a human-centred view of the universe four centuries after the Copernican revolution broke up the enclosed relationship between humankind, the cosmos and the creator.

Almost as striking as interstellar photography is the way in which certain figures or diagrams have acquired a resonance and authority that is almost that of an icon. The Hertzsprung-Russell diagram, developed between 1908 and 1913, which relates stars' spectral type to luminosity encapsulates some of the basics of astrophysics by showing effectively the life history of stars. The placing of any star on the diagram is like placing one more feature on the map of cosmic space-time. Less technical but equally momentous is the small picture-mosaic constructed to show cosmic background radiation, the echoes of the universe's birth, which has been widely taken as proving the Big Bang theory of cosmology. When this background radiation was first detected, one scientist described it as the 'handwriting of God' on the universe. Perhaps the ultimate cosmological diagrams are those which set out to plot the distribution of galaxies in the universe. The prototype of these shows the familiar, comforting pattern of celestial longitude around its edge, while the galaxies radiate out across 600,000 light years. Like the traditional planisphere, this is a view from above the north celestial pole, but here only a limited latitudinal cross-section is shown. The distinctive pattern at this declination (27–32 degrees) has been christened the 'Stick Man': so we use familiar images to focus and interpret something almost inconceivable, and make it amenable to human thought. These are the new maps of the heavens, and this small diagram is perhaps the most ambitious star chart ever drawn.

The authority accorded to these cosmic images sets up the most compelling echoes of the theme with which this book began – the designation of star-patterns in the ancient Near East. Their creators needed landmarks in the sky, as we do. Our sky is no longer their sky, but we still need to elucidate its mysteries by visualising patterns contained within it. The sun-god, the bull in the heavens, the morning star, the Pleiades – all these fed the astral religions of ancient cultures. Today our secular mysticism is sustained by radiant images of galaxies evolving in the vastness of space. Such an image is both an answer to our questions and a further question: it shows how much we have learned about the cosmos, and yet how much remains still unexplained. The path towards the unknown is always marked by elements of the known. The map of the heavens is evolving into unexpected forms, but its shaping force has familiar roots deep in the human psyche: the eternal desire to answer the questions, 'Where do we come from? What are we? Where are we going?' To chart the universe of which we are a part, and to place ourselves within it, is the question that comes before and after all the other questions that humankind ever asks. It is curious to reflect that the discoveries of modern science have made that place more mysterious to us than it was to the people of the ancient civilisations with which this book began.

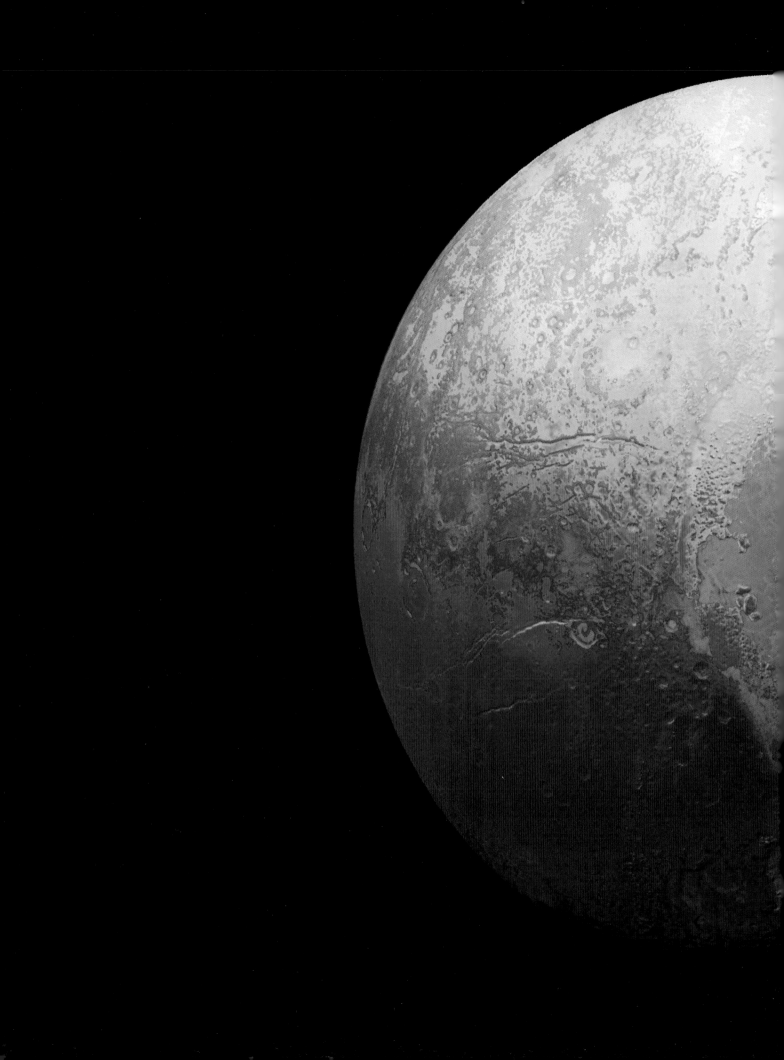

Further Reading

Aratus' *Phaenomena*, translated by D. B. Gain, London, 1976.

Aristotle: *Physics and On the Heavens*, translated by R. Hardie, R. Gaye, & J. Stocks, Oxford, 1922.

K. Brecher & M. Feirtag (eds): *The Astronomy of the Ancients*, Cambridge, 1979.

Copernicus: *On the Revolutions of the Heavenly Spheres*, translated by E. Rosen, New York, 1978.

Vincenzo Coronelli: *Libro dei Globi,* 1701, facsimile edition, Amsterdam, 1969.

Dante: *Paradiso: Illuminations to Dante's Divine Comedy*, edited by J. Pope-Hennessy, London, 1993.

E. Dekker: *Illustrating the Phaenomena: Celestial Cartography in Antiquity and the Middle Ages,* Oxford, 2013.

W. Derham: *Astro-Theology*, London, 1715 (and many subsequent editions).

M. Faintich: *Astronomical Symbols on Ancient and Medieval Coins*, North Carolina, 2008.

C. Flammarion: *Astronomie Populaire*, 1880, English ed. by J. Gore, London, 1895.

P. Grego: *Blazing a Ghostly Trail: Ison and Great Comets of the Past*, New York, 2014.

J. B. Harley & D. Woodward (eds): *History of Cartography*, Vol. 1, Chicago, 1987.

J. B. Harley & D. Woodward (eds): *History of Cartography*, Vol. 2, Part 1, Chicago, 1992.

D. B. Hermann: *History of Astronomy from Herschel to Hertzsprung*, English ed., Cambridge, 1984.

N. Kanas: *Star Maps : History, Artistry and Cartography,* New York, 2012.

D. Malin: *A View of the Universe*, Cambridge, 1993.

M. Milanesi: *Vincenzo Coronelli, Cosmographer*, Turnhout, 2016.

O. Neugebauer: *Astronomy and History*, New York, 1983.

O. Neugebauer & R. Parker: *Egyptian Astronomical Texts*, 4 vols. Providence, RI, 1960–69.

J. D. North: *Fontana History of Astronomy and Cosmology*, London, 1994.

J. D. North: *The Measure of the Universe: A History of Modern Cosmology*, New York, 1990.

J. D. North: *Chaucer's Universe*, Oxford, 1990.

O. Pedersen: *Early Physics and Astronomy*, new ed., Cambridge, 1993.

Plato: *Timaeus and Critias*, translated by D. Lee, Harmondsworth, 1965.

Ptolemy: *Almagest*, English translation by G. J. Toomer, London, 1984.

A. Sandage: *The Hubble Atlas of Galaxies*, Washington, 1963.

G. M. Sesti: *The Glorious Constellations*, New York, 1991.

H. Shapley & H. Howarth: *A Source Book in Astronomy*, Cambridge, MA, 1929.

G. S. Snyder: *Maps of the Heavens*, London, 1984.

A. Turner: *Early Scientific Instruments, Europe 1400–1800*, London, 1987.

A. Van Helden: *Measuring the Universe*, Chicago, 1985.

C. Walker (ed.): *Astronomy Before the Telescope*, London, 1996.

D. Warner: *The Sky Explored*, Amsterdam, 1979.

Index

Page numbers in *italics* indicate illustrations or maps; the suffix *t* indicates information in a table; the suffix *c* indicates information in a caption.